**Decision and Control
in Uncertain Resource Systems**

This is Volume 172 in
MATHEMATICS IN SCIENCE AND ENGINEERING
A Series of Monographs and Textbooks
Edited by RICHARD BELLMAN, *University of Southern California*

The complete listing of books in this series is available from the Publisher upon request.

Decision and Control in Uncertain Resource Systems

MARC MANGEL

Department of Mathematics
University of California, Davis
Davis, California

ACADEMIC PRESS, INC.
(Harcourt Brace Jovanovich, Publishers)

Orlando   San Diego   New York   London
Toronto   Montreal   Sydney   Tokyo

COPYRIGHT © 1985, BY ACADEMIC PRESS, INC.
ALL RIGHTS RESERVED.
NO PART OF THIS PUBLICATION MAY BE REPRODUCED OR
TRANSMITTED IN ANY FORM OR BY ANY MEANS, ELECTRONIC
OR MECHANICAL, INCLUDING PHOTOCOPY, RECORDING, OR ANY
INFORMATION STORAGE AND RETRIEVAL SYSTEM, WITHOUT
PERMISSION IN WRITING FROM THE PUBLISHER.

ACADEMIC PRESS, INC.
Orlando, Florida 32887

United Kingdom Edition published by
ACADEMIC PRESS, INC. (LONDON) LTD.
24/28 Oval Road, London NW1 7DX

Library of Congress Cataloging in Publication Data

Mangel, Marc.
  Decision and control in uncertain resource systems.

  (Mathematics in science and engineering)
  Includes bibliographical references and index.
  1. Stochastic processes.   2. Control theory.
3. Dynamic programming.   4. Natural resources.   5. Renewable natural resources.   I. Title.   II. Series.
QA274.M35   1983   519.2   83-22333
ISBN 0-12-468720-2

PRINTED IN THE UNITED STATES OF AMERICA

85 86 87 88   9 8 7 6 5 4 3 2 1

Dedicated to the triumvirate Marsha Barber, Susan Milke Mangel, and Kris Ryan, who are directly responsible, and to the triumvirate Susan, Jennifer, and Bethann, who are indirectly responsible

Contents

Preface   ix
Acknowledgments   xiii
Introduction   1

1. Discrete and Continuous Stochastic Processes
1.1. Probability   6
1.2. The Brownian Motion Process $W(t)$ 17
1.3. The Jump Process $\pi(t)$ 22
1.4. Mean and Variance Propagation in Stochastic Differential Equations   27
1.5. Kalman Filtering and Its Extensions   33

2. Stochastic Control and Dynamic Programming
2.1. The Principle of Optimality   42
2.2. The Dynamic Programming Equation for a $W(t)$-Driven Process   44
2.3. The Dynamic Programming Equation for a $\pi(t)$-Driven Process   50
2.4. The Dynamic Programming Equation with Parameter Uncertainty   53
2.5. Finding Optimal Stochastic Controls Directly   58
2.6. Control with Probability Criteria   61
2.7. Myopic Bayes Stochastic Problems and Nonlinear Programming   65
2.8. Values, Strategies, and Policy Iteration   68

3. Key Results in the Deterministic Theory of Resource Exploitation
3.1. Classification of Resource Problems and the Concept of Utility   74
3.2. Exploitation of an Exhaustible Resource: Optimal Extraction Policy and Price Dynamics   79
3.3. Exploitation of a Renewable Resource: Optimal Harvest Rates and Price Dynamics   85
     Appendix A: A Review of the Calculus of Variations and Optimal Control Theory   87

4. Exploration for and Optimal Utilization of Exhaustible Resources

4.1. Exploration for Exhaustible Resources 94
4.2. Utilization of an Uncertain Resource without Learning 107
4.3. Utilization of an Uncertain Resource with Learning 111
4.4. Optimal Exploration for and Exploitation of an Exhaustible Resource 121
4.5. Price Dynamics and Markets for Exhaustible Resources 127

5. Exploration, Observation, and Assessment of Renewable Resource Stocks

5.1. Introduction to the Theory of Line Transects 132
5.2. Surveys of Fish Stocks 138
5.3. Model Identification for Aggregating Fisheries 148
5.4. Search Effort and Estimates of Stock Size 158

6. Management of Fluctuating Renewable Resource Stocks

6.1. The Allocation of Fishing Rights 170
6.2. Harvesting a Randomly Fluctuating Population 175
6.3. Dealing with Parameter Uncertainty in Managed Populations 188
6.4. Price Dynamics in Renewable Resource Markets 194
6.5. Optimal Trade-off between Economic Return and the Risk of Undesirable Events 198

7. Management of Mixed Renewable and Exhaustible Resource Systems: Agricultural Pest Control

7.1. The Operational Formulation: Cotton in California 204
7.2. Models for Pest Population Dynamics 206
7.3. Models with Population Genetics 210
7.4. The Single-Season Economic Optimization Problem 214
7.5. The Multiseason Economic Optimization Problem 218

8. Introduction to Numerical Techniques

8.1. Newton's Method 221
8.2. Sequence and Series Acceleration Techniques 226
8.3. Solving Stochastic Differential Equations Numerically 228
8.4. Some Numerical Considerations for DPEs 234

References 239

Index 251

Preface

> All the business of war, and indeed all the business of life, is to find out what you don't know by what you do; that's what I called "guessing what was on the other side of the hill."
>
> The Duke of Wellington

This book evolved from a number of graduate courses in resource management and operations research. The courses had the common theme that the problems of interest are dynamic with unknown and/or stochastic components. The last course, Mathematics of Renewable Resources, was designed as a research-level literature survey course. A draft of the manuscript for this book was used in this course. The draft was given to the students in advance of the lectures and was then revised after classroom experience and comments.

The theme of this book is resource exploitation under uncertainty. Exploration as well as exploitation is considered throughout. In general, the exploration side of resource management has not received the consideration that it deserves. The best way to learn about the nature of resource stocks is to go out and search for them, to find out what you don't know by what you do.

While this book can be used by researchers as a sourcebook for the literature, I view it as a graduate-level textbook. In particular, there are many students who are interested in resource management involving uncertainty but do not know where to begin their study. This volume should provide a means to reach the subject with a reasonable amount of study and should build the confidence needed to read the literature and work in the area. As a graduate-level text, this book is intended to survey a vast intellectual landscape. A broad range of topics is covered, but none of the topics in exceedingly great detail (as one would find in a research monograph, for example). The reference list should provide the reader who wants more detail a starting point. Indeed, I hope that a reader who gets through the book will emerge with the ability and courage to charge off into the literature.

The mathematical requirements for reading the book are introductory courses in probability or statistics and a good knowledge of calculus. A course in deterministic models for resource management, at the level of Clark (1976) or Dasgupta and Heal (1979), is undoubtedly helpful, but not essential.

The book is as self-contained as possible; and, wherever feasible, simple methods are used instead of complex ones. For example, in Chapter 4 there is a problem in which the Kuhn–Tucker theorem could be used to give a quick (but nonintuitive) answer. There is, however, a method of obtaining the solution that uses calculus and common sense instead, and this method is the one that is used (it also allows a much deeper interpretation of the solution). In Chapter 5 the Kuhn–Tucker theorem is really needed, so it is introduced there.

The book begins with two chapters on mathematics. Chapter 1 contains a review of probability theory and an introduction to the stochastic processes that are needed in the rest of the book. The notation is set in this chapter; whenever possible, probabilistic or stochastic concepts are illustrated with models from resource management. The ideas of Bayesian analysis are introduced here, since they provide a cornerstone for the rest of the material. Chapter 2 contains a discussion of optimal stochastic control, which is a major tool used in the rest of the book. In particular, the methods of stochastic dynamic programming are discussed in detail. Examples from resource management are used to illustrate these ideas. Each section in the first two chapters contains a list of sections in the rest of the book where the methods are used. These are not easy mathematical techniques, but they are useful and they answer important questions.

The next five chapters contain specific resource applications of the methods. Chapter 3 is a review of the deterministic models used in resource management. It serves the dual purpose of introducing the appropriate notation for the rest of the book and framing the problems of interest. This chapter also contains a review of optimal control theory and the calculus of variations. Chapter 4 is concerned with the optimal exploration for and exploitation of exhaustible resources. Most of the ideas in Chapters 1–3 are used in some way in Chapter 4. Chapters 5 and 6 discuss the optimal use of renewable resources. Chapter 5 is solely concerned with the exploration and assessment of renewable resource stocks. Chapter 5, and the relevant sections of Chapter 4, contain an introduction to search theory as it is applied to problems in ecology and resource management. Chapter 6 is devoted to the optimal exploitation of renewable resource stocks, using mainly the methods of stochastic dynamic programming.

Chapter 7 is a quasi-deterministic interlude. It concerns the management of a "mixed" resource system, namely, a system in which one is simultaneously

PREFACE   xi

dealing with a renewable and an exhaustible resource stock. The problem treated in Chapter 7 is that of agricultural pest control, but many of the ideas apply to other similar problems.

Chapter 8 is concerned with numerical methods that are useful for stochastic control problems. Once again, the object of this chapter is to provide the reader with a knowledge of some common techniques, the jargon, and sufficient sense of the methods that he or she can understand detailed methods in numerical analysis.

This book could be used in either a lecture course or a seminar course. It contains enough material for a quick-paced two-quarter lecture course or a slower-paced full-year lecture course. The last time that I taught this material, I was able to cover Sections 1.1–1.4 and 2.1–2.4, Chapter 3, Sections 4.1–4.5, 5.1–5.4, and 6.1–6.5, Chapter 7, and Chapter 8. Although I did it, I felt rushed. There are two general outlines that one can use. The first route starts with Chapter 1 and covers the mathematical methods before the applications. The second route starts with Chapter 3 and goes back to Chapters 1 and 2 as necessary.

There is sufficient material here for a year-long advanced seminar course in which the students present different sections. Either of the two routes would work well.

Acknowledgments

I especially thank the graduate students at the University of California at Davis who followed the entire set of courses from which this book evolved. They are Jim Anderson, Larry Flynn, Larry Karp, and Spiro Stefanou. They constantly provided stimulation and challenge to me and never accepted anything because the professor said it was so.

I have had the good luck to work with a number of outstanding scientists who helped to mold my viewpoint. I especially thank Colin W. Clark and Donald Lugwig at the University of British Columbia, Phil DePoy at the Center for Naval Analyses, and Bill Hay at the University of Colorado.

The original work that appears in this book was supported in part by a contract from the Office of Naval Research (NK00014-18-K-0030) and a grant from the National Science Foundation (MCS-81-21659). I thank both organizations for their support of my work and the NSF for continuing patronage.

Marla Milner did an excellent and speedy job of typing (and retyping) the manuscript.

Introduction

The two decades between 1962 and 1982 saw an incredible increase in the interest of many individuals and groups in the management of natural resources. Rachel Carson's book "Silent Spring," published in 1962, warned of the environmental hazards and damage that can be caused by technological development. The OPEC oil embargo of 1973–1974 highlighted the crucial dependence of the developed nations on certain important natural resource stocks. The near extinction of a number of species of whales and the spectacular collapse of fish stocks, for example, the Peruvian Anchovy, showed the vulnerability of natural resource stocks to the pressures introduced by mankind through harvesting. The protection of dolphins associated with schools of tuna, through the Marine Mammal Protection Act, showed the complicated impact of biological management policy on the economic return obtained from a different source. These examples illustrate the breadth, depth, and complexity of the issues associated with natural resource management. Study of these examples also shows that analysis has a valuable role in the proper management of natural resources.

The use of analytical methods in resource management can be traced to at least 1931, when Harold Hotelling published a landmark paper on the economics of exhaustible resources. In this paper Hotelling applied the techniques of the calculus of variations to determine the optimal extraction and consumption patterns for exhaustible resource stocks. The paper of Devarajan and Fisher (1981) contains an excellent review of the influence that Hotelling's work has had in the last 50 years. Major steps in applying these sorts of ideas to renewable resource stocks can be traced to Gordon (1954) and Schaefer (1957). Gordon's paper dealt with the economic analysis of common property resources, mainly fisheries. Gordon clearly pointed out the externality of the common property resource and the dissipation of economic return. That is, an individual fisherman has no motivation for conserving the stock, since other fishermen will harvest it. Schaefer explicitly introduced a model for the study of the interaction between the population dynamics of tuna and the economic dynamics of tuna fishermen. (In this book we shall see that tuna harvesting still provides a wealth of problems and ideas.) Much of the work started by Hotelling, Gordon, and Schaefer is now called *mathematical resource economics* (for exhaustible resource stocks) or *mathematical bioeconomics* (for renewable resource stocks). Another major breakthrough occurred when Clark was able to do for renewable resource problems, in a very simple fashion, what Hotelling did for exhaustible resource problems. These areas are quite popular right now, especially when deterministic models are used. Clark's (1976) book summarizes much of the work on and spurred interest in such problems.

Note as well known as any of the above-mentioned works, but equally as interesting (especially in the light of historial events), is a National Academy of Sciences Report published in 1963. This report, written by D. Bronk, D. Frasche, K. Hubbert, P. Morse, F. Notestein, R. Revelle, and P. Weiss, addresses the conditions of resource stocks in the United States. The report is described in some detail on pages 339–342 of Morse's autobiography (Morse, 1977). The first paragraph of the report is worth reading:

> The United States now possesses or has access to the gross quantities of energy, food, fiber, minerals, water and space required to maintain and even improve its standard of living for perhaps four to five decades, if the world situation were to remain relatively stable. But the world situation is not stable; other countries may preempt what we now import and new demands may exhaust our own resources more rapidly than we now foresee. We must aquire understanding or our alternatives and take concerted action if we are to maintain our present access without future denial to our children of the freedom of action we now enjoy. We live in a critical period when clear definition of the natural resources research objectives and the development plans for realizing them are of great importance to the future welfare and stability of mankind. Today the United States is able to enjoy a certain complacency concerning its short-range situation; but it faces a wider arena and a longer term in which complacency would be tragically dangerous.

The report contained the following ten recommendations concerning research in natural resources.

(1) Extend applied research to increase productivity in a wider range of agricultural environments.

(2) Conduct basic research in plant and animal genetics and breeding.

(3) Support research leading to low-cost sources of industrial energy.

(4) Increase support of research in the movement and quality of surface water and groundwater.

(5) Develop analytic techniques for the planning and management of water resources.

(6) Increase support for research relevant to the discovery and development of mineral deposits.

INTRODUCTION

(7) Conduct research on pollution and its effects on man's total environment.

(8) Support ocean fisheries research.

(9) Develop systems analysis capabilities for resource planning and management.

(10) Establish a Central Natural Resources Group within the federal government to coordinate, support, and extend this planning and research.

This list contains the outline of a broad research program in resource management. The techniques and methods of analysis developed in this book touch on points 1, 5, 6, 8, and 9 in this list.

The development of mathematical bioeconomics and resource economics was rapid. In general, deterministic models are used to study the problems of interest. This is done for two reasons. First, in general the models are simpler to deal with than nondeterministic models. Second, fewer parameters are needed. It is becoming increasingly clear, however, that when dealing with natural resource systems one must consider the uncertainty in parameter values, the large fluctuations that occur in nature, and the lack of information about such systems. For management purposes new models are needed rather than descriptive paradigms. One approach is to take an already existing deterministic model and add "noise" to it. This approach is often used in mathematical biology [e.g., May (1973)] and clearly results in hard mathematical and analytical problems. The usefulness and meaning of such ad hoc models is always open to question. Another approach is to look at the components of the system of interest and to model them separately, trying to sort out where the important uncertainties are located. Both approaches are used in this book; the reader should make his own decision regarding the utility of each kind of approach.

The importance of uncertainty in economics was recognized very early by Frank Knight in his classic book "Risk, Uncertainty, and Profit" (Knight, 1971; a reprint of the original version published in 1921). On page 268, he wrote:

> With the introduction of uncertainty—the fact of ignorance and necessity of acting upon opinion rather than knowledge—into this Eden-like situation, its character is completely changed. With uncertainty absent, man's energies are devoted altogether to doing things; it is doubtful whether intelligence itself would exist in such a situation... With uncertainty present, doing things, the actual execution of activity, becomes in a real sense a secondary part of life; the primary problem or function is deciding what to do and how to do it.

This book is concerned with exactly that function—deciding what to do and how to do it.

Problems in which parameters are unknown or uncertain and in which state variables fluctuate rarely have exact solutions. Heuristic and approximate methods are needed. One heuristic commonly used in engineering and economic problems is the certainty equivalent (CE) approach. According to this method, random variables associated with a particular problem are replaced by their average values. This procedure converts a stochastic problem into a deterministic one. When system dynamics are nonlinear or probability distributions have many local maxima, the CE approach may produce results of limited or no usefulness.

Another approach to these problems is the "systems engineering" viewpoint. Sage (1977) describes it this way: "Systems engineering is an appropriate combination of the mathematical theory of systems and behavioral theory in a useful setting appropriate for the resolution of real world problems." To see how such a view might be applied, consider the following canonical model in fisheries analysis:

$$dX/dt = g(X(t)) - qEX(t).$$

Here $X(t)$ is the biomass of the fish stock at time t, $g(X)$ a natural growth rate, E the fishing effort, and q a catchability coefficient. A stochastic term can be added to the right-hand side to try to model fluctuations. Then one must deal with very difficult questions in nonlinear stochastic differential equations. Statistical questions concern, for example, how one finds an estimate for q.

The systems approach leads to the following kind of question: What does fishing effort mean? To answer this question, one must study the process of fishing and the behavior and motivation of fishermen. Then one observes that the search for fish is an important part of the fishing process. (There is even some humor present. *Theorem*: fishermen must find fish. *Proof*: by contradiction.) Search, as a topic of analysis, is generally ignored in most books on resource systems analysis. An exception, for instance, is the book edited by Smith and Krutilla (1982) that contains one chapter on exploration.

A successful study of resource management must involve the mathematical, statistical, and systems engineering viewpoints in the appropriate combination.

In addition to finding resource stocks, assessing the size of the stock is extremely important in many problems. Information about stock size may, for example, drive the solution to an entire complex problem. This book contains two chapters (4 and 5) with sections dedicated to questions of search and exploration.

It should be clear that to analyze the resource problems of interest here, one needs more tools than differential equations and a smattering of economics. Probability and statistics must become common tools for researchers working on resource problems. Often individuals knowledgeable in deterministic areas hesitate to use probability, stochastic processes, and statistics. It is hoped that this book can contribute to a transition for them.

A general comment about the role of analysis in management problems is also appropriate. The analyst is rarely the policymaker. Any tractable model is only a caricature of reality and should not be construed as an exact representation of reality (or, worse yet, confused as one). The policymaker has many considerations to take into account in addition to the results of analysis. Hence, the optimal role for the analyst is to provide results that can be used as a decision aid by the policymaker. The author's nonacademic work and consulting experience has been as an analyst who acts as a constultant for the decisionmaker. This is, to some extent, the view advocated in the book.

In this book the author has tried to be comprehensive in that most resource areas are discussed at some level. There are two glaring omissions. The first is water resource systems analysis. It is not included here because there are a number of excellent books on the subject, for example, those by Haimes (1977), Kottegoda (1980), and Loucks, Stedinger, and Haith (1981). (A final exam question for one of the courses that lead to this book asked the students to determine a good problem from water resource management and to describe possible solutions.) The second omission is the general area of bidding, leases, options, and futures markets. This topic is not included, because it deserves a separate book which uses, perhaps, one-third of the material in this book but then treats the general question of the economics of uncertainty and information. A nice introduction to this problem area is the book by Smiley (1979), which treats the problem of bidding for offshore oil leases. The paper by Engelbrecht-Wiggans (1980) contains a good introduction to the literature on auctions and bidding.

1

Discrete and Continuous Stochastic Processes

The main tool for the analysis of resource systems with uncertainty is the theory of stochastic processes. In this chapter we review this theory and present a few new results with resource system applications. In each section of the chapter there are a few key references so that a reader who chooses to can review further. The goal of this review is to provide a strong foundation in the probabilistic tools that will be needed later on, rather than a comprehensive treatment.

1.1 PROBABILITY

This section on probability is essential to the results that follow throughout the book. The stochastic processes are characterized by a random variable $X(t)$ on some probability space Ω (Ω remains in the background), which is characterized by its distribution function $F(x, t)$:

$$\Pr\{X(t) \leq x\} = F(x, t). \tag{1.1.1}$$

If the derivative $f(x, t) = \partial F(x, t)/\partial x$ exists, it can be used to characterize $X(t)$ as follows:

$$\begin{aligned}\Pr\{x \leq X(t) \leq x + dx\} &= F(x + dx, t) - F(x, t) \\ &= \left\{F(x, t) + \frac{\partial F}{\partial x}(x, t)\, dx + o(dx)\right\} - F(x, t) \\ &= f(x, t)\, dx + o(dx). \end{aligned} \tag{1.1.2}$$

1.1. PROBABILITY

The second step in this calculation uses a Taylor expansion of $F(x + dx, t)$ around (x, t). Here $o(dx)$ indicates terms that are of higher order than dx, that is, a term is $o(dx)$ if $\lim_{dx \to 0} o(dx)/dx = 0$. The function $f(x, t)$ is called the *density* of $X(t)$.

Example 1.1.1: The Normal Density

One choice for $F(x, t)$ is the normal distribution

$$F(x, t) = (1/\sqrt{2\pi}) \int_{-\infty}^{x/\sigma\sqrt{t}} \exp(-s^2/2) \, ds, \tag{1.1.3}$$

with corresponding density

$$f(x, t) = (1/\sqrt{2\pi t \sigma^2}) \exp(-x^2/2\sigma^2 t). \tag{1.1.4}$$

For the distribution (1.1.3) and the density (1.1.4), we write $X(t) \sim N(0, \sigma^2 t)$. As t decreases to zero, the density (1.1.4) becomes smaller and smaller if $x \neq 0$ but increases at $x = 0$. In fact, as $t \to 0$, $f(x, t) \to \delta(x)$, the Dirac delta function [see Lighthill (1975) for an excellent introduction to such generalized functions]. That is, $\delta(x)$ is zero everywhere except at the origin, and if $g(x)$ is a regular function, then $\int \delta(x) g(x) \, dx = g(0)$. We shall find a number of uses for $\delta(x)$.

The variable $X(t)$ could have a discrete distribution, in which case one writes

$$\Pr\{X(t) = k\} = p_k(t), \quad k = 0, 1, 2, \ldots. \tag{1.1.5}$$

Let $g(X(t))$ be a function of $X(t)$ and define

$$E\{g(X(t))\} = \begin{cases} \sum_{k=0}^{\infty} g(k) p_k(t) & \text{if } X \text{ is discrete,} \\ \int f(x, t) g(x) \, dx & \text{if } X \text{ is continuous.} \end{cases} \tag{1.1.6}$$

The quantity on the left-hand side of (1.1.6) is the expectation of $g(X(t))$. In particular, two important quantities are the mean $E\{X(t)\}$ and variance $\text{Var}\{X(t)\}$ defined by

$$E\{X(t)\} \equiv \bar{X}(t),$$

$$\text{Var}\{X(t)\} \equiv \sigma^2(t) = E\{X^2(t)\} - E\{X(t)\}^2.$$

A nondimensional measure of the spread of a random variable is the coefficient of variation defined by

$$CV(X(t)) = \sigma(t)/\bar{X}(t).$$

Another useful fact to know is that if $X(t) \geq 0$, then

$$E\{X(t)\} = \int_0^\infty (1 - F(x, t)) \, dx.$$

A quick proof of this result is obtained using integration by parts.

Often, one is interested in more than one random variable. For a pair $X(t)$, $Y(t)$, the joint distribution function is defined by

$$\Pr\{X(t) \leq x, Y(t) \leq y\} = F(x, y, t). \tag{1.1.7}$$

The random variables $X(t)$ and $Y(t)$ are independent if

$$F(x, y, t) = \left[\lim_{y \to \infty} F(x, y, t)\right]\left[\lim_{x \to \infty} F(x, y, t)\right]. \tag{1.1.8}$$

The limits in (1.1.8) are the *marginal* distribution functions. If it exists, the joint density function $f(x, y, t)$ satisfies

$$F(x, y, t) = \int_{-\infty}^x \int_{-\infty}^y f(u, v, t) \, dv \, du. \tag{1.1.9}$$

The marginal densities are defined by

$$f_X(x, t) = \int_{-\infty}^\infty f(x, v, t) \, dv,$$

$$f_Y(y, t) = \int_{-\infty}^\infty f(u, y, t) \, du. \tag{1.1.10}$$

If $X(t)$ and $Y(t)$ are independent, then $f(x, y, t) = f_X(x, t) f_Y(y, t)$.

Let X and Y have joint density $f(x, y)$ (with t suppressed). Set $Z = X + Y$. Then the density $f_Z(z)$ is given by

$$f_Z(z) = \int_{-\infty}^\infty f(x, z - x) \, dx. \tag{1.1.11}$$

If X and Y are independent, one obtains

$$f_Z(z) = \int_{-\infty}^\infty f_X(x) f_Y(z - x) \, dx; \tag{1.1.12}$$

this integral is a convolution.

Example 1.1.2

Suppose that X and Y are independent with *gamma densities* $\gamma(v_1, \alpha_1)$ and $\gamma(v_2, \alpha_2)$, respectively. The densities are

$$f_X(x) = e^{-\alpha_1 x} \alpha_1^{v_1} x^{v_1 - 1}/\Gamma(v_1),$$

$$f_Y(y) = e^{-\alpha_2 y} \alpha_2^{v_2} y^{v_2 - 1}/\Gamma(v_2), \tag{1.1.13}$$

1.1. PROBABILITY

for $x \geq 0$, $y \geq 0$. Here $\Gamma(v) = (v-1)!$ is the gamma function (Abramowitz and Stegun, 1964). If $Z = X + Y$, the density for Z is [from (1.1.12)]

$$f_Z(z) = \frac{\alpha_1^{v_1}\alpha_2^{v_2}}{\Gamma(v_1)\Gamma(v_2)} \int_0^z e^{-\alpha_1 x} x^{v_1-1} e^{-\alpha_2(z-x)} (z-x)^{v_2-1}\, dx \qquad (1.1.14)$$

$$= \frac{\alpha_1^{v_1}\alpha_2^{v_2} e^{-\alpha_2 z}}{\Gamma(v_1)\Gamma(v_2)} \int_0^z e^{-\alpha_1 x} x^{v_1-1} e^{\alpha_2 x}(z-x)^{v_2-1}\, dx. \qquad (1.1.15)$$

Now assume that $\alpha_1 = \alpha_2 = \alpha$, so that

$$f_Z(z) = \frac{\alpha^{v_1+v_2} e^{-\alpha z}}{\Gamma(v_1)\Gamma(v_2)} \int_0^z x^{v_1-1}(z-x)^{v_2-1}\, dx. \qquad (1.1.16)$$

Setting $x = zu$ gives

$$f_Z(z) = \alpha^{v_1+v_2} e^{-\alpha z} z^{v_1+v_2-1} \frac{\int_0^1 u^{v_1-1}(1-u)^{v_2-1}\, du}{\Gamma(v_1)\Gamma(v_2)}. \qquad (1.1.17)$$

From a book on special functions [e.g., Abramowitz and Stegun (1964)],

$$\frac{\int_0^1 u^{v_1-1}(1-u)^{v_2-1}\, du}{\Gamma(v_1)\Gamma(v_2)} = \frac{1}{\Gamma(v_1+v_2)}. \qquad (1.1.18)$$

Thus the density for Z is

$$f_Z(z) = \frac{\alpha^{v_1+v_2} e^{-\alpha z} z^{v_1+v_2-1}}{\Gamma(v_1+v_2)}, \qquad z \geq 0, \qquad (1.1.19)$$

another gamma density.

Exercise 1.1.3

Suppose that X_i, $i = 1, 2$, are normally distributed with mean μ_i and variance σ_i^2. Then the density for X_i is

$$f_{X_i}(x_i) = \frac{1}{\sqrt{2\pi}\,\sigma_i} \exp\left\{\frac{-(x_i - \mu_i)^2}{2\sigma_i^2}\right\}. \qquad (1.1.20)$$

Show that $X_1 + X_2$ is normally distributed with mean $\mu_1 + \mu_2$ and variance $\sigma_1^2 + \sigma_2^2$.

Exercise 1.1.4

Suppose that X_i, $i = 1, 2$, are normally distributed with mean 0 and variance σ_i^2.

(1) Show that X_i^2 has a gamma density $\gamma(1/2, 1/2\sigma_i^2)$.
(2) Find a representation of the density of $X_1^2 + X_2^2$.

In addition to the sum, one often needs to find the quotient of two random variables. If X and Y have joint density $f(x, y)$ and $Z = Y/X$, then the density for Z is given by

$$f_Z(z) = \int |x| f(x, xz)\, dx. \tag{1.1.21}$$

Exercise 1.1.5

Suppose that X and Y are independent with gamma densities $\gamma(v_1, \alpha)$ and $\gamma(v_2, \alpha)$, respectively. Show that the density of $Z = Y/X$ is

$$f_Z(z) = \begin{cases} 0, & z \leq 0, \\ \dfrac{\Gamma(v_1 + v_2)}{\Gamma(v_1)\Gamma(v_2)} \dfrac{z^{v_2 - 1}}{(z + 1)^{v_1 + v_2}}, & z > 0. \end{cases}$$

Often, the questions encountered in this book involve the probability of one event, given that another event occurred. This leads to the concept of conditional probability. Define the conditional probability of A given B by (assuming $\Pr\{B\} > 0$)

$$\Pr\{A|B\} = \Pr\{A, B\}/\Pr\{B\}. \tag{1.1.22}$$

In this equation $\Pr\{A|B\}$ is the probability of event A occurring, given that event B occurred and $\Pr\{A, B\}$ is the joint probability that both A and B occur. The idea behind (1.1.22) is best illustrated with a diagram, as shown in Fig. 1.1.1. The hatched region represents $\Pr\{A, B\}$ (assuming that the area of Ω is normalized to 1). Equation (1.1.22) can be rewritten to give

$$\Pr\{A, B\} = \Pr\{A|B\}\Pr\{B\}.$$

Interchanging A and B on the right-hand side gives

$$\Pr\{A, B\} = \Pr\{B|A\}\Pr\{A\}.$$

Consequently, another form of (1.1.22) is

$$\Pr\{A|B\} = \Pr\{B|A\}\Pr\{A\}/\Pr\{B\}. \tag{1.1.22'}$$

Equation (1.1.22'), and its various versions, is known as the *Bayes theorem*. It provides a way to use information obtained through observations of the stochastic process to update estimates of probabilities. When (1.1.22) is rewritten in terms of densities, it becomes

$$f(x|y) = f(x, y)/f_Y(y) \quad \text{if } f_Y(y) \neq 0 \tag{1.1.23}$$

1.1. PROBABILITY

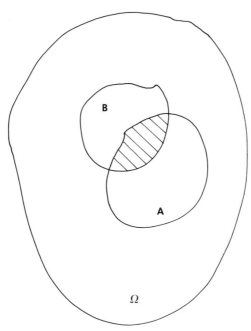

Fig. 1.1.1 An illustration of the concept of conditional probability. Here Ω is the set of all possible events and A and B are two subsets. The shaded region is the set of points where both A and B occur.

(with the obvious extension for discrete variables). The conditional expectation is defined by

$$E\{g(X)|y\} = \begin{cases} \int g(x) f(x,y) \, dx / f_Y(y) & \text{if } X, Y \text{ are continuous,} \\ \sum g(k) \Pr\{X = k, Y = j\} / \Pr\{Y = j\} & \text{if } X, Y \text{ are discrete.} \end{cases}$$
(1.1.24)

Furthermore, applying the definition of conditional probability gives

$$E\{g(X)h(Y)\} = \int E\{g(X)|Y = y\} h(y) f_Y(y) \, dy. \qquad (1.1.25)$$

When $h(y) = 1$, (1.1.25) is known as the law of total probabilities:

$$E\{g(X)\} = \begin{cases} \int E\{g(X)|Y = y\} f_Y(y) \, dy & \text{if } X, Y \text{ are continuous,} \\ \sum_j E\{g(X)|Y = j\} \Pr\{Y = j\} & \text{if } X, Y \text{ are discrete.} \end{cases}$$
(1.1.26)

Rewriting (1.1.23) gives

$$f(x, y) = f_Y(y)f(x|y), \qquad (1.1.27)$$

which leads to the density form of Bayes theorem:

$$f(x|y) = f_X(x)f(y|x) \Big/ \int f_X(x)f(y|x)\,dx. \qquad (1.1.28)$$

Bayes theorem is a tool that is used over and over again in this book. The law of total probabilities is illustrated in the next example.

Example 1.1.6

Suppose that the number of schools of fish in a region, $\tilde{N}$, has a Poisson distribution, so that

$$\Pr\{\tilde{N} = k|\lambda\} = e^{-\lambda}(\lambda^k/k!). \qquad (1.1.29)$$

Using the Poisson model (1.1.29) is equivalent to assuming that the schools of fish are randomly distributed [see, for example, Pielou (1977) for a discussion of such assumptions]. If the region of interest is large enough, it may happen that λ itself is unknown. Then (1.1.29) is viewed as a conditional distribution, and the unconditional one is found by integrating (1.1.29) against the distribution on λ. For example, assume that λ has a gamma distribution with parameters v, α:

$$\Pr\{\lambda \in (\bar{\lambda}, \bar{\lambda} + d\bar{\lambda})\} = \frac{\exp(-\alpha\bar{\lambda})\bar{\lambda}^{v-1}}{\Gamma(v)}\alpha^v\,d\bar{\lambda}. \qquad (1.1.30)$$

Using the law of total probabilities gives the unconditional distribution for $\tilde{N}$:

$$\Pr\{\tilde{N} = k\} = \int_0^\infty \frac{\exp(-\lambda)\lambda^k}{k!} \frac{\exp(-\alpha\lambda)\lambda^{v-1}}{\Gamma(v)}\alpha^v\,d\lambda, \qquad (1.1.31)$$

or

$$\Pr\{\tilde{N} = k\} = \frac{1}{k!}\frac{\alpha^v}{(\alpha+1)^{v+k}}\frac{\Gamma(v+k)}{\Gamma(v)}. \qquad (1.1.32)$$

Equation (1.1.32) can be rewritten as

$$\Pr\{\tilde{N} = k\} = \binom{k+v-1}{k}\left(\frac{\alpha}{\alpha+1}\right)^v\left(\frac{1}{\alpha+1}\right)^k, \qquad (1.1.33)$$

which is the negative binomial distribution. The moments of $\tilde{N}$ are

$$E\{\tilde{N}\} = v/\alpha, \quad \text{Var}\{\tilde{N}\} = v/\alpha + (1/v)(v/\alpha)^2. \qquad (1.1.34)$$

1.1. PROBABILITY

This distribution is often used by theoretical ecologists to represent patchiness. Note that if v and α increase in such a way that v/α is constant but $v \to \infty$, then (1.1.33) gives back the Poisson process. Thus v is sometimes called a measure of patchiness.

The next example shows how Bayes theorem can be used.

Example 1.1.7

Assume that the setup of Example 1.1.6 is changed so that the number of schools of fish discovered in $(0, t)$ is a Poisson random variable with parameter λt.

Assume that in a time period of length t, M schools of fish were found. Then

$$\Pr\{\lambda \in (\bar{\lambda}, \bar{\lambda} + d\bar{\lambda}) | M\} = \Pr\{\lambda \in (\bar{\lambda}, \bar{\lambda} + d\bar{\lambda}), M\}/\Pr\{M\}. \quad (1.1.35)$$

The numerator in equation (1.1.35) can be found by multiplying the conditional probability of finding M schools of fish in $(0, t)$ given λ by the density for λ. That is,

$$\Pr\{\lambda \in (\bar{\lambda}, \bar{\lambda} + d\bar{\lambda}), M\} = \frac{e^{-\bar{\lambda}t}(\bar{\lambda}t)^M}{M!} \frac{e^{-\alpha\bar{\lambda}}\bar{\lambda}^{v-1}\alpha^v}{\Gamma(v)}.$$

The denominator in (1.1.35) is the numerator integrated over λ:

$$\int_0^\infty e^{-\alpha\lambda}\frac{\lambda^{v-1}}{\Gamma(v)}\alpha^v e^{-\lambda t}\frac{(\lambda t)^M\,d\lambda}{M!} = \frac{\alpha^v t^M}{\Gamma(v)M!}\frac{\Gamma(v+M)}{(\alpha+t)^{v+M}}. \quad (1.1.36)$$

The posterior (after the discovery) probability then becomes

$$\Pr\{\lambda \in (\bar{\lambda}, \bar{\lambda} + d\bar{\lambda}) | M\} = \frac{\exp[-(\alpha+t)\bar{\lambda}]\bar{\lambda}^{v+M-1}(\alpha+t)^{v+M}}{\Gamma(v+M)}, \quad (1.1.37)$$

which is another gamma density with parameters $v + M$ and $\alpha + t$. Thus, the form of the density on λ does not change.

This next example shows how much care must be exercised when using the concepts of conditional probability. (A fuller discussion of the basic idea in this example is found in Section 2.8.)

Example 1.1.8

Assume that X_1 and X_2 are independent random variables with the same density function $f(x, \tilde{\mu})$, where $\tilde{\mu}$ is an unknown parameter. Assume that $g(\mu, \bar{\mu})$ is the density function for $\tilde{\mu}$.

The problem is to find the density for $Y = X_1 + X_2$. The correct method proceeds as follows:

$$\Pr\{Y \in (y, y + dy)\} = \int \Pr\{Y \in (y, y + dy)|\mu\} \Pr\{\tilde{\mu} \in (\mu, \mu + d\mu)\} \quad (1.1.38)$$

$$= \int \Pr\{X_1 + X_2 \in (y, y + dy)|\mu\} g(\mu, \bar{\mu}) \, d\mu$$

$$= \left[\int \left[\int f(y_1, \mu) f(y - y_1, \mu) \, dy_1 \right] g(\mu, \bar{\mu}) \, d\mu \right] dy. \quad (1.1.39)$$

Observe that the conditional distributions were used until the very last step of the calculation. With the incorrect method one finds unconditional distributions too early in the calculation. Observe that

$$\Pr\{X_i \in (x, x + dx)\} = \int f(x, \mu) g(\mu, \bar{\mu}) \, d\mu \, dx. \quad (1.1.40)$$

In the incorrect method, one proceeds as follows:

$$\Pr\{Y \in (y, y + dy)\} = \Pr\{X_1 + X_2 \in (y, y + dy)\}$$

$$= \int \Pr\{X_1 \in (y_1, y_1 + dy_1)\} \Pr\{X_2 \in (y - y_1, y - y_1 + dy)\}$$

$$= \left[\int \left[\int f(y_1, \mu) g(\mu, \bar{\mu}) \, d\mu \right] [f(y - y_1, \mu) g(\mu, \bar{\mu}) \, d\mu] \, dy_1 \right] dy.$$

$$(1.1.41)$$

It is easy to see that (1.1.39) and (1.1.41) are not the same. When using the incorrect method, one forces the X_i to take the same distribution before the calculation is done.

Exercise 1.1.9

(i) Reexamine equations (1.1.39) and (1.1.40). Determine the conditions on $g(\mu, \bar{\mu})$ that would make the two equations the same.

(ii) Assume that X_i are normally distributed with unknown mean μ and variance σ^2. Assume that μ is normally distributed with mean $\bar{\mu}$ and variance σ^2. Find the densities of Y using both methods. Before doing the calculation, decide which density should have a larger variance and why.

The last tools of this section are the characteristic function

$$\varphi(t) = \mathrm{E}\{e^{itX}\} \quad (1.1.42)$$

1.1. PROBABILITY

and generating function
$$g(S) = E\{S^X\}. \tag{1.1.43}$$

Since (1.1.42) and (1.1.43) are the same if $S = e^{it}$, only the generating function will be discussed.

Recall the following properties of generating functions.

(1) The generating function of a sum of indepedent random variables is the product of the generating functions.

(2) The moments of X are given by
$$E\{X(X-1)\cdots(X-k)\} = (d^{(k+1)}g(S)/dS^{k+1})|_{S=1}.$$

Example 1.1.10

The generating function can be used to prove a useful identity. Suppose that X_i are independent, identically distributed random variables and that
$$Y = \sum_{i=1}^{N} X_i \tag{1.1.44}$$

is a random sum of X_i, where N is a random variable independent of the X_i. Then the first two moments of Y are given by
$$\begin{aligned} E\{Y\} &= E\{X_i\} E\{N\}, \\ \text{Var}\{Y\} &= E\{X_i\}^2 \sigma_N^2 + E\{N\} \sigma_{X_i}^2. \end{aligned} \tag{1.1.45}$$

To show this, form the generating function of Y:
$$\begin{aligned} g_Y(S) &= E\{S^Y\} = E\{S^{X_1 + \cdots + X_N}\} \\ &= E\{E\{S^{X_1 + \cdots + X_N} | N\}\} \\ &= \sum_{n=0}^{\infty} E\{S^{X_1 + \cdots + X_N} | N = n\} \Pr\{N = n\}. \end{aligned}$$

Since the X_i are independent of each other and of N, $g_Y(S)$ can be rewritten as
$$\begin{aligned} g_Y(S) &= \sum_{n=0}^{\infty} E\{S^{X_1}\} E\{S^{X_2}\} \cdots E\{S^{X_n}\} \Pr\{N = n\} \\ &= \sum_{n=0}^{\infty} g(S)^n \Pr\{N = n\} = E\{g(S)^N\} = g_N(g(S)). \end{aligned} \tag{1.1.46}$$

Thus
$$E\{Y\} = dg_N(g(S))/dS|_{S=1} = g_N'(g(S))g'(S)|_{S=1}, \tag{1.1.47}$$

which leads to the first formula in (1.1.45). The reader should verify the second one as an exercise.

Example 1.1.11

Suppose that there is a region of size A which is sampled for schools of fish and that an area of size a is sampled. If there are N schools in A, one expects the number of schools found, X, to satisfy (for randomly distributed schools) the binomial distribution

$$\Pr\{X = k | N\} = \binom{N}{k}\left(\frac{a}{A}\right)^k\left(1 - \frac{a}{A}\right)^{N-k}. \quad (1.1.48)$$

To understand (1.1.48), observe that a/A is the probability that any one of the N schools will be caught, if the schools are randomly distributed. Thus, a/A is a success probability. Equation (1.1.48) describes the probability of k successes in N trials. Now assume that N itself has a Poisson distribution. What can one say about the distribution of X?

The number X is generated by summing up N Bernoulli (0 or 1) random variables, where the probability of a success is a/A. The generating function for these Bernoulli variables is $f(s) = (1 - a/A)s^0 + (a/A)s$. The generating function of the Poisson process is

$$g_N(s) = \exp(-\lambda + \lambda s) \quad (1.1.49)$$

(prove this). The generating function of X becomes

$$\begin{aligned} g_X(s) = g_N(f(s)) &= \exp[-\lambda + \lambda f(s)] \\ &= \exp\{-\lambda + \lambda[(1 - a/A) + (a/A)s]\} \\ &= \exp[(-\lambda a/A) + (\lambda a/A)s]. \end{aligned} \quad (1.1.50)$$

Equation (1.1.50) shows that X is itself a Poisson process with parameter $\lambda a/A$.

Bibliographic Notes

Good introductions to probability theory are the books by Feller (1968), Hoel, Port, and Stone (1971), Karlin and Taylor (1977), and Ross (1980). Analytical tools from calculus, such as the Taylor expansion and integration by parts, will be used throughout this book and should be reviewed if needed. A standard reference on the use of the negative binomial distribution to represent patchiness is the book by Pielou (1977).

1.2 THE BROWNIAN MOTION PROCESS $W(t)$

This section and the next characterize two important stochastic processes. The basic idea is this: when time is measured continuously, which it is in both sections, then the extremes of behavior for the state variable in a stochastic process are completely continuous and completely discontinuous paths. The first kind of process is considered in this section. The second kind (jump processes) is considered in the next section.

Brownian motion is a continuous time- and state-space stochastic process. It is used in the following sections: 2.1, on dynamic programming; 2.4, on finding optimal controls directly; 2.5, on control with probability criteria; 4.5, on price dynamics in exhaustable resource markets; 6.2, on harvesting a randomly fluctuating population; 6.4, on the trade-off between stock and control fluctuations; and 6.5, on price dynamics in renewable resource markets.

Brownian motion, $W(t)$, is a stochastic process with the following properties.

(1) Sample paths of $W(t)$ are continuous.
(2) $W(0) = 0$.
(3) The increment $W(t + \tau) - W(\tau)$ is normally distributed with mean 0 and variance $\sigma^2 t$.
(4) If (t, τ) and $(\hat{t}, \hat{\tau})$ are disjoint intervals, then $W(\tau) - W(t)$ and $W(\hat{\tau}) - W(\hat{t})$ are independent random variables.

Setting $dW = W(t + dt) - W(t)$, property (2) is written as

$$\Pr\{\eta \leq dW \leq \eta + d\eta\} = (1/\sqrt{2\pi\sigma^2 \, dt}) \exp(-\eta^2/2\sigma^2 \, dt) \, d\eta. \quad (1.2.1)$$

The first two moments of dW are

$$E\{dW\} = 0, \qquad E\{(dW)^2\} = \sigma^2 \, dt. \quad (1.2.2)$$

The fact that the variance of dW is of order dt creates many mathematical difficulties. For example,

$$E\{(dW/dt)^2\} \sim \sigma^2/dt \to \infty \quad \text{as} \quad dt \to 0, \quad (1.2.3)$$

so that $W(t)$ does not have a derivative in a strict mathematical sense. Thus, Brownian motion is everywhere continuous and nowhere differentiable.

Another feature of (1.2.2) arises in Taylor expansions. Consider a function $X = f(t, W(t))$. The consistent Taylor expansion of X is then

$$dX = \frac{\partial f}{\partial t} dt + \frac{\partial f}{\partial W} dW + \frac{1}{2}\left\{\frac{\partial^2 f}{\partial t^2}(dt)^2 + 2\frac{\partial^2 f}{\partial t \, dW} dt \, dW + \frac{\partial^2 f}{\partial W^2}(dW)^2\right\}$$
$$+ O((dt)^3) + O((dW)^3). \quad (1.2.4)$$

Here $O((dt)^3)$ indicates a term that is a constant times $(dt)^3$. Now, X is really a random variable, so that (1.2.4) only makes sense in terms of moments or distributions. Taking the expectation of (1.2.4) gives

$$E\{dX \mid W(t) = w\} = \frac{\partial f(t, w)}{\partial t} dt + \frac{1}{2} \frac{\partial^2 f(t, \omega)}{\partial W^2} \sigma^2 \, dt + o(dt).$$

An extra term appears, caused by the fact that $E\{(dW)^2\} \propto dt$. This leads to a new calculus (the Ito calculus) in which one symbolically writes

$$dX = \frac{\partial f}{\partial t} dt + \frac{\partial f}{\partial W} dW + \frac{1}{2} \frac{\partial^2 f}{\partial W^2} dt + o(dt), \tag{1.2.5}$$

where $o(dt)/dt \to 0$ as $dt \to 0$.

Also, note that if $t < s$ [since $W(t) = W(t) - W(0)$],

$$E\{W(t)W(s)\} = E[\{W(s) - W(t)\}W(t)] + E\{W(t)^2\}$$
$$= E\{(W(s) - W(t))W(t)\} + \sigma^2 t = \sigma^2 t, \tag{1.2.6}$$

and if $t > s$, $E\{W(t)W(s)\} = \sigma^2 s$. Thus the correlation function of $W(t)$ is

$$E\{W(t)W(s)\} = \sigma^2 \min(t, s). \tag{1.2.7}$$

Equation (1.2.3) indicates that dW/dt does not exist (in the mathematical sense). This has nothing to do with physical reality; rather, it is caused by assumption (4). Namely, if two disjoint intervals are close enough, there must be some correlation between the two variables. This point, which actually has little effect on the applications in this book, is discussed by Wong (1971).

The derivative $\xi(t) = dW/dt$ is extremely useful. It is called *Gaussian white noise* (GWN). The correlation function of GWN is

$$E\{\xi(t)\xi(s)\} = \frac{\partial^2}{\partial t \, \partial s} E\{W(t)W(s)\} = \frac{\partial^2}{\partial t \, \partial s} \min(t, s)\sigma^2 = \delta(t - s)\sigma^2. \tag{1.2.8}$$

To verify (1.2.8), proceed as follows. First, observe that

$$\frac{\partial}{\partial t} \min(t, s) = \begin{cases} 1, & t < s, \\ 0, & t > s, \end{cases}$$

which is a step function, taking the step at $t = s$. This step function can be approximated by $(1/\sqrt{2\pi}) \int_{n(t-s)}^{\infty} \exp(-y^2/2) \, dy$, where n is taken to be a large parameter. With this approximation,

$$\frac{\partial^2}{\partial t \, \partial s} \min(t, s) = \lim_{n \to \infty} \frac{\partial}{\partial s} \frac{1}{\sqrt{2\pi}} \int_{n(t-s)}^{\infty} \exp\left(\frac{-y^2}{2}\right) dy$$

$$= \lim_{n \to \infty} \frac{n}{\sqrt{2\pi}} \exp\left[\frac{-n^2(t-s)^2}{2}\right] = \delta(t-s).$$

1.2. THE BROWNIAN MOTION PROCESS $W(t)$

The reader should verify that $\lim_{n\to\infty} (n/\sqrt{2\pi}) \exp[-n^2(t-s)^2/2]$ is indeed the delta function (see Example 1.1.1).

Properties of Brownian motion are easily derived from the transition density, which is defined through assumption (2) to be

$$q(x, t; y, \tau) \, dx = \Pr\{W(t) \in (x, x + dx) | W(\tau) = y\}$$
$$= \frac{1}{\sqrt{2\pi\sigma^2(t-\tau)}} \exp\left\{-\frac{(x-y)^2}{2\sigma^2(t-\tau)}\right\} dx. \quad (1.2.9)$$

In (1.2.9) it is assumed that $t > \tau$.

Exercise 1.2.1

Show that $q = q(x, t; y, \tau)$ in (1.2.9) satisfies

$$\frac{\partial q}{\partial t} = \frac{\sigma^2}{2} \frac{\partial^2 q}{\partial x^2}, \quad \frac{\partial q}{\partial \tau} + \frac{\sigma^2}{2} \frac{\partial^2 q}{\partial y^2} = 0. \quad (1.2.10)$$

Equations (1.2.10) are diffusion equations.

From the transition density, one obtains properties of the *sample path*. For example, if $0 < t_1 < t_2 < \cdots < t_n$, then

$$\Pr\{W(t_i) \in (x_i, x_i + dx_i), i = 1, \cdots, n\}$$
$$= q(x_1, t_1, 0, 0) q(x_2, t_2; x_1, t_1)$$
$$\cdots q(x_n, t_n; x_{n-1}, t_{n-1}) \, dx_1 \cdots dx_n. \quad (1.2.11)$$

Exercise 1.2.2

Show that if $0 \leq t < 1$, then

$$\Pr\{W(t) \in (x, x + dx), W(1) \in (y, y + dy)\}$$
$$= \frac{1}{2\pi\sigma\sqrt{t(1-t)}} \exp\left\{-\frac{1}{2}\left(\frac{x^2}{t} + \frac{(y-x)^2}{1-t}\right)\right\} dx \, dy. \quad (1.2.12)$$

Exercise 1.2.3

Are the following functions also Brownian motion?

$$W_1(t) = \alpha W(t/\alpha^2), \quad W_2(t) = t W(1/t). \quad (1.2.13)$$

The natural extensions of Brownian motion are *diffusion processes*, which provide the underpinnings for an entire theory of *stochastic differential equations* (SDE). Instead of specifying a transition density, as in (1.2.9), one specifies the moments of the density. In particular, if $X(t)$ is the process of interest and $q(x, t, y, \tau) = \Pr\{X(t) \in (x, x+dx) | X(\tau) = y\}$, then assume

$$\int q(y + \xi, \tau + dt, \xi, \tau) y \, dy = b(\xi, \tau) \, dt + o(dt),$$

$$\int q(y + \xi, \tau + dt, \xi, \tau) y^2 \, dy = a(\xi, \tau) \, dt + o(dt), \qquad (1.2.14)$$

$$\int q(y + \xi, \tau + dt, \xi, \tau) y^n \, dy = o(dt), \qquad n \geq 3.$$

In addition to (1.2.14), it is assumed that the paths of $X(t)$ are continuous. If $X(t)$ denotes the diffusion process and $dX \equiv X(t + dt) - X(t)$, then (1.2.14) is equivalent to

$$E\{dX | X(t) = x\} = b(x, t) \, dt + o(dt),$$

$$E\{(dX)^2 | X(t) = x\} = a(x, t) \, dt + o(dt), \qquad (1.2.15)$$

$$E\{(dX)^n | X(t) = x\} = o(dt), \qquad n \geq 3.$$

The law of total probability gives the *Chapman–Kolmogorov* relation:

$$q(x, t, \xi, \tau) \, dx = \int q(\xi + y, \tau + dt, \xi, \tau) q(x, t, \xi + y, \tau + dt) \, dy \, dx. \quad (1.2.16)$$

[*Optional exercise*: Restate (1.2.16) verbally to ensure that you understand it.] A Taylor expansion of (1.2.16) gives

$$q(x, t, \xi, \tau) \, dx = \int [q(x, t, \xi, \tau) + q_\tau(x, t, \xi, y) \, dt + O((dt)^2)$$

$$+ y q_\xi(x, t, \xi, \tau) + \tfrac{1}{2} y^2 q_{\xi\xi}(x, t, \xi, \tau) + O(dt y^2) + O(y^3)]$$

$$\times q(\xi + y, \tau + dt, \xi, \tau) \, dy \, dx. \qquad (1.2.17)$$

In this equation subscripts indicate partial derivatives. Applying (1.2.14) to (1.2.17) shows that, to leading order in dt,

$$0 = dt\{q_\tau + b(\xi, \tau) q_\xi + \tfrac{1}{2} a(\xi, \tau) q_{\xi\xi}\} \, dt + o(dt). \qquad (1.2.18)$$

Dividing (1.2.18) by dt and letting $dt \to 0$ shows that $q(x, t, \xi, \tau)$ satisfies the diffusion equation

$$q_\tau + b(\xi, \tau) q_\xi + \tfrac{1}{2} a(\xi, \tau) q_{\xi\xi} = 0. \qquad (1.2.19)$$

The variables ξ, τ are called "backward" variables, since they relate to the starting position and time of the process. The variables x, t are called "forward" variables. Equation (1.2.19) is called the Kolmogorov backward equation.

1.2. THE BROWNIAN MOTION PROCESS $W(t)$

It can be shown that in the forward variables x and t, q satisfies (Feller, 1971) the Kolmogorov forward or Fokker–Planck equation

$$\frac{\partial q}{\partial t} = \frac{1}{2}\frac{\partial^2}{\partial x^2}(aq) - \frac{\partial}{\partial x}(bq). \quad (1.2.20)$$

Equations (1.2.19) and (1.2.20) are diffusion equations; hence the name "diffusion processes."

A more intuitive, though less rigorous, method for deriving equations for diffusion processes is illustrated by the following example. Let $X(t)$ be the diffusion process defined by (1.2.15) with $a(\cdot)$ and $b(\cdot)$ independent of time and consider the average time $T(x)$ that it takes $X(t)$ to exit the interval (l_1, l_2), given that $X(0) = x$, $l_1 < x < l_2$. Now, in time dt, the process jumps to a new point $x + dX$. One relates $T(x)$ to the expected time starting at this new point by the expression

$$T(x) = dt + \mathrm{E}_{dX}\{T(x + dX)\}. \quad (1.2.21)$$

Equation (1.2.21) is obtained through a simple use of conditional probability. The expectation is taken over all possible jumps dX. A Taylor expansion of (1.2.21) gives

$$0 = dt + \mathrm{E}_{dX}\{T'(x)\,dX + (\tfrac{1}{2}T''(x))(dX)^2 + O((dX)^3)\}. \quad (1.2.22)$$

Applying (1.2.15), dividing by dt, and letting $dt \to 0$ gives

$$-1 = b(x)T'(x) + \tfrac{1}{2}a(x)T''(x). \quad (1.2.23)$$

Exercise 1.2.4

What boundary conditions go with (1.2.23)?

Exercise 1.2.5

Assume $u(x, t) = \Pr\{X(t) \text{ has exited } (l_1, l_2) \text{ through } l_2 \text{ by time } t \mid X(0) = x\}$. Derive the diffusion equation for $u(x, t)$ and determine the appropriate boundary and initial conditions.

Bibliographic Notes

Good general discussions on Brownian motion can be found in the books by Karlin and Taylor (1977), Ludwig (1974), and Schuss (1980). The Ito calculus is explained by Schuss (1980) and in "cookbook" form by Kamien and Schwartz (1981). Schuss (1980) and Wong (1971) provide a good general introduction to stochastic differential equations.

1.3 THE JUMP PROCESS $\pi(t)$

Brownian motion is characterized by a continuum of changes. At the other extreme is the jump process, in which the changes in the state variable occur in discrete units. The jump process is used in Sections 2.2 and 2.3 on dynamic programming, 4.1 on exploration for exhaustible resources, 4.3 on utilization of an exhaustible resource with learning, 4.4 on optimal exploration and exploitation of an exhaustible resource, 5.3 on surveys of fish stocks, and 5.5 on search effort and estimates of predicted harvest.

The general jump process $\pi(t)$ is characterized by the following assumptions.

(1) $\pi(0) = 0$.
(2) Given that $\pi(t) = x$, the waiting time until the next jump has an exponential distribution with expectation $1/\alpha(x)$.
(3) Given that a jump occurs, if $\Delta\pi$ is the change in π, then

$$\Pr\{\Delta\pi = \delta_1\} = \lambda(x), \qquad \Pr\{\Delta\pi = -\delta_2\} = \mu(x). \qquad (1.3.1)$$

where δ_1 and δ_2 are positive.
(4) The jumps are independent of the waiting times.

These three assumptions lead to a semi-Markov process. This means that the evolution of the process is Markovian (where it goes depends only on its current state), but the exit time from the current state depends on the state. In particular, assumption (2) shows that the probability of exit from $\pi(0) = x$ by time t is $1 - e^{-\alpha(x)t}$.

In this book the most important jump processes are the ones in which $\alpha(x) \equiv 1$, so that the postulates change to

$$\Pr\{\pi(t+dt) - \pi(t) = \delta_1 | \pi(t) = x\} = \lambda(x)\,dt + o(dt),$$
$$\Pr\{\pi(t+dt) - \pi(t) = -\delta_2 | \pi(t) = x\} = \mu(x)\,dt + o(dt),$$
$$\Pr\{\pi(t+dt) - \pi(t) = 0 | \pi(t) = x\} = 1 - [\lambda(x) + \mu(x)]\,dt + o(dt). \qquad (1.3.2)$$
$$\Pr\{\text{any other transition} | \pi(t) = x\} = o(dt)$$

Example 1.3.1

For the Poisson process $\pi(0) = 0$, $\delta_1 = 1$, $\lambda(x) = \lambda$ (a constant), and $\mu(x) = 0$. The assumptions then reduce to

$$\Pr\{\pi(t+dt) - \pi(t) = 1\} = \lambda\,dt + o(dt),$$
$$\Pr\{\text{no change}\} = 1 - \lambda\,dt + o(dt). \qquad (1.3.3)$$

1.3. THE JUMP PROCESS $\pi(t)$

This gives, after iteration,

$$\Pr\{\pi(t) = j\} = e^{-\lambda t}(\lambda t)^j/j!, \quad (1.3.4)$$

which is the standard result for the Poisson process.

Equations (1.3.1) and (1.3.2) can be combined by setting $d\pi = \pi(t + dt) - \pi(t)$. Then (1.3.1) and (1.3.2) are written in a form similar to (1.2.15) as

$$E\{d\pi | \pi(t) = x\} = [\lambda(x)\delta_1(x) - \mu(x)\delta_2(x)] \, dt + o(dt), \quad (1.3.5)$$

$$E\{(d\pi)^2 | \pi(t) = x\} = [\lambda(x)\delta_1^2(x) + \mu(x)\delta_2^2(x)] \, dt + o(dt). \quad (1.3.6)$$

When $\delta_1 = \delta_2 = 1$, the analysis can go much further. If $\pi(0) = 0$, $\pi(t)$ can only take integer values. Let λ_n, and μ_n denote $\lambda(n)$ and $\mu(n)$, where n is an integer, and define $P_n(t)$ by

$$P_n(t) = \Pr\{\pi(t) = n\}. \quad (1.3.7)$$

The basic assumptions show that $P_n(t + dt)$ is related to $P_n(t)$ by

$$P_n(t + dt) = P_n(t)\{1 - (\lambda_n + \mu_n) \, dt\} \\ + (\lambda_{n-1} \, dt)P_{n-1}(t) + (\mu_{n+1} \, dt)P_{n+1}(t) + o(dt). \quad (1.3.8)$$

In (1.3.8) it is understood that $P_k \equiv 0$ if $k < 0$. Subtracting $P_n(t)$ from both sides, dividing by dt, and letting $dt \to 0$ gives the set of ordinary differential equations

$$dP_n/dt = -(\lambda_n + \mu_n)P_n(t) + \lambda_{n-1}P_{n-1}(t) + \mu_{n+1}P_{n+1}(t). \quad (1.3.9)$$

The initial conditions for (1.3.8) are derived by observing that $\pi(0) = 0$ is the same as

$$P_0(0) = 1, \quad P_n(0) = 0, \quad n \geq 1. \quad (1.3.10)$$

Exercise 1.3.2

Derive (1.3.9) from (1.3.8). What is special about $P_0(t)$?

The *generating function* $f(t, s)$ for $\pi(t)$ is defined by

$$f(t, s) = \sum_n P_n(t)s^n. \quad (1.3.11)$$

Exercise 1.3.3

In a pure birth process, for which

$$\lambda_n = n\lambda, \quad \mu_n = 0, \quad (1.3.12)$$

show that $f(t, s)$ satisfies the partial differential equation

$$\frac{\partial f}{\partial t} = -\lambda s \frac{\partial f}{\partial s} + \lambda s^2 \frac{\partial f}{\partial s} \tag{1.3.13}$$

and solve it.

Exercise 1.3.4

Suppose that

$$\lambda_n = n\lambda, \qquad \mu_n = n\mu. \tag{1.3.14}$$

Show that $f(t, s)$ satisfies

$$\frac{\partial f}{\partial t} + \{(\lambda + \mu)s - \lambda s^2 - \mu\} \frac{\partial f}{\partial s} = 0 \tag{1.3.15}$$

and solve it.

Example 1.3.5

As an example of such processes, suppose that there are M schools of fish in a region and that the search for fish is characterized by the following rule:

Pr{discovering a school in $(t, t + dt)|n$ schools were discovered by time t}

$$= \begin{cases} \lambda(1 - n/M) \, dt + o(dt), & n \leq M - 1, \\ 0, & \text{otherwise.} \end{cases} \tag{1.3.16}$$

In (1.3.16), λ is a measure of the rate at which schools of fish are found. The equations for $P_n(t)$ are found to be

$$\frac{dP_n}{dt} = -\lambda P_n + \lambda P_{n-1} - \left\{ \frac{\lambda(n-1)}{M} P_{n-1} - \frac{\lambda n}{M} P_n \right\}, \quad 1 \leq n \leq M - 1,$$

$$\frac{dP_0}{dt} = -\lambda P_0, \tag{1.3.17}$$

$$P_0(0) = 1, \qquad P_n(0) = 0, \quad n \geq 1. \tag{1.3.18}$$

One can verify that

$$P_n(t) = \binom{M}{n}(1 - \exp(-\lambda t/M))^n \exp[-\lambda t(M - n)/M] \tag{1.3.19}$$

is a solution of (1.3.18). (Are there other solutions? Why or why not?) Equation (1.3.19) shows that the number of schools discovered in $(0, t)$ is binomial with parameters M and $1 - e^{-\lambda t/M}$.

1.3. THE JUMP PROCESS $\pi(t)$

Note that the *expected* number of schools found is, from (1.3.19), $M \times [1 - \exp(-\lambda t/M)]$. If $M \to \infty$, $M[1 - \exp(-\lambda t/M)] \sim \lambda t + O(1/M)$, so that the Poisson process average is obtained as a limit. In fact, if $M \to \infty$, then (1.3.16) reduces to the Poisson assumption.

The equation for the generating function for this process is

$$\frac{\partial f}{\partial t} + \frac{\lambda}{M}(s^2 - s)\frac{\partial f}{\partial s} = \lambda(s-1)f \tag{1.3.20}$$

(verify it as an exercise).

Exercise 1.3.6

As $M \to \infty$, $f(t, s) \to f_p(t, s)$, which satisfies (from 1.3.20)

$$\partial f_p/\partial t = \lambda(s-1)f_p. \tag{1.3.21}$$

Show that $f_p(t, s)$ is the generating function of the Poisson process.

Equation (1.3.20) and the other equations for generating functions are first-order partial differential equations. Such equations can be solved quite effectively by the method of characteristics (John, 1978). To illustrate this method, (1.3.20) will now be solved explicitly.

To begin the calculation, set

$$ds/dt = (\lambda/M)(s^2 - s). \tag{1.3.22}$$

Integrating (1.3.22) gives

$$s - 1 = ske^{(\lambda/M)t}, \tag{1.3.23}$$

where k is a constant of integration. The key to finding a solution of (1.3.20) is the observation that, along the solution curves of (1.3.22), the left-hand side of (1.3.20) becomes the total derivative of $f(t, s)$. Hence, along these curves one obtains the ordinary differential equation for f:

$$df/dt = \lambda(s-1)f. \tag{1.3.24}$$

Solving (1.3.23) for s, inserting into (1.3.24), and integrating gives

$$\int_0^t [s(\tau) - 1]\, d\tau = \int_0^t ke^{(\lambda/M)\tau}/(1 - ke^{\lambda\tau/M})\, d\tau$$

$$= -(M/\lambda)\ln[(1 - ke^{\lambda t/M})/(1 - k)]. \tag{1.3.25}$$

Thus $f(t)$ is explicitly given by

$$f = \exp\{-M \ln[(1 - ke^{\lambda t/M})/(1 - k)]\}. \tag{1.3.26}$$

Substituting for s gives

$$f(t, s) = \{s - (s - 1)e^{-\lambda t/M}\}^M. \tag{1.3.27}$$

Equation (1.3.27) gives the generating function explicitly. Taking the derivative in (1.3.27) and setting $s = 1$ gives

$$E\{\text{number of schools discovered in } (0, t)\} = M(1 - e^{-\lambda t/M}), \tag{1.3.28}$$

which is the result obtained from (1.3.19) directly.

To see how such a simple model could be used, one can add uncertainty about λ and some simple economics to the problem. First assume that λ is not known exactly but has a gamma distribution with parameters v, α. Assume also that k vessels are in the region and they work independently, so that λ in (1.3.16) is replaced by $k\lambda$. Then, instead of (1.3.28), one has

$$E\{\text{number of schools discovered by } k \text{ vessels in } (0, t)|\lambda\} = M(1 - e^{-k\lambda t/M}). \tag{1.3.29}$$

Since λ is unknown, (1.3.29) should be averaged over λ. When this is done, one obtains (using the gamma distribution) an expected catch of

$$E\{\text{number of schools discovered by } k \text{ vessels in } (0, t)\}$$
$$= M[1 - \alpha^v/(\alpha + kt/M)^v]. \tag{1.3.30}$$

To include some economic aspects, assume that each school is worth p dollars and the cost of operating is c_1 dollars fixed cost and c_2 dollars per unit time. The net gain of the fishery in $(0, t)$ is

$$G(t, k) = pM[1 - (1 + kt/\alpha M)^{-v}] - c_1 k - c_2 kt. \tag{1.3.31}$$

Assume that fishing stops at a time t_s such that the marginal return vanishes, that is, when

$$\partial G/\partial t|_{t_s} = 0. \tag{1.3.32}$$

(There could be other choices of t_s, e.g., income target levels or seasonal closures. These are discussed in chapters 3 and 6.) The value of t_s obtained from (1.3.32) depends on k. The total return at time t_s is $G(t_s(k), k)$. This return can be maximized over k to find the optimal number of vessels that should participate in the fishery. Results of such calculations are shown in Table 1.3.1. This simple example shows how far a little bit of analysis can go to help one understand an interesting problem. [This example, however, is somewhat contrived. First, there is no way to incorporate learning about the value of λ (see Example 2.4.1 for a treatment of learning). Second, the interpretation of λ in (1.3.16) is not completely clear. More realistic examples come later in the book.]

1.4. MEAN AND VARIANCE PROPAGATION IN SDEs

Table 1.3.1

Optimal Gains in a Fishery with Depletion[a]

v	α	k^*	Gain ($)
$E\{\lambda\} = 0.7$			
1.4	2.0	4	5426
0.7	1.0	3	4014
0.467	0.667	3	3153
0.35	0.50	2	2638
$E\{\lambda\} = 1.4$			
0.7	0.5	4	9909
0.350	0.25	3	6736
0.233	0.167	2	5135
0.175	0.125	2	4083

[a] Parameter values: $c_1 = \$2000$, $c_2 = \$400$, $p = \$4000$, $M = 10$.

Bibliographic Notes

Feller (1968) and Ludwig (1974) contain good discussions on the jump process. The Ludwig book, in particular, stresses many biological applications and provides examples of solutions of the equations for generating functions. The method of characteristics is ably described by John (1978). Another good (the classic) sourcebook is that by Courant and Hilbert (1962). It will be a rare occasion in this book that ordinary differential equations such as (1.3.22) will have explicit solutions. Good reference books for ordinary differential equations are Boyce and DiPrima (1977) at the introductory level and Coddington and Levinson (1955) at the advanced level.

1.4 MEAN AND VARIANCE PROPAGATION IN STOCHASTIC DIFFERENTIAL EQUATIONS

Many, if not all, of the interesting equations in this book are nonlinear equations with random and uncertain components. Thus, exact solutions of the equations of interest will rarely be found. Instead, various approximation methods are needed. One class of approximations involves questions about the moments of the stochastic process. That is, suppose that $X(t)$ satisfies a stochastic differential equation and the explicit solution of such an equation cannot be found. One can ask if it is possible to obtain information about the moments of $X(t)$, say, the mean and variance, and this information to help characterize the behavior of the process. The process $X(t)$ satisfies a stochastic differential equation (SDE) if its increment $dX = X(t + dt) - X(t)$ is given by

$$dx = b(t, X) \, dt + \sqrt{a(t, X)} \, dW \tag{1.4.1}$$

or

$$dx = b(t, X) \, dt + a(t, X) \, d\pi, \tag{1.4.2}$$

where $b(t, X)$ and $a(t, X)$ are given functions. If $a(t, X) \equiv 0$, then (1.4.1) and (1.4.2) reduce to the ordinary differential equation $dX/dt = b(t, X)$. This equation has a deterministic solution when $a(t, X)$ is zero, but the solution $X(t)$ of (1.4.1) or (1.4.2) is otherwise a random process. This section contains a description of methods that yield the mean and variance of $X(t)$ directly from (1.4.1) or (1.4.2). These methods will be called *mean–variance algorithms*. The results in this section are used in Section 4.5 on the price dynamics of an exhaustible resource, 6.2 on harvesting a randomly fluctuating population, 6.5 on the price dynamics of a renewable resource, and 8.3 on numerical methods.

Equation (1.4.1) and (1.4.2) are interpreted as follows. For (1.4.1), assume that $X(t) = x$. Then $dX = X(t + dt) - X(t)$ is normally distributed with mean $b(t, x) \, dt + o(dt)$ and variance $a(t, x)\sigma^2 \, dt + o(dt)$, where

$$E(dW^2) = \sigma^2(t) \, dt.$$

For (1.4.2), if $X(t) = x$, then dX has a distribution such that

$$\begin{aligned}
\Pr\{dX = b(t, x) \, dt + a(t, x)\delta_1(x)\} &= \lambda(x) \, dt + o(dt), \\
\Pr\{dX = b(t, x) \, dt - a(t, x)\delta_2(x)\} &= \mu(x) \, dt + o(dt), \\
\Pr\{dX = b(t, x) \, dt\} &= 1 - (\lambda(x) + \mu(x)) \, dt + o(dt), \\
\Pr\{\text{any other value for } dX\} &= o(dt).
\end{aligned} \tag{1.4.3}$$

Associated with (1.4.1, 1.4.2) are an initial distribution on $X(0)$ and the densities for both $X(0)$ and $X(t)$:

$$\begin{aligned}
\Pr\{X(0) \in (x_0, x_0 + dx_0)\} &= p_0(x_0) \, dx_0, \\
\Pr\{X(t) \in (x, x + dx)\} &= p(x, t) \, dx.
\end{aligned} \tag{1.4.4}$$

In terms of the densities in (1.4.4), the mean $\bar{X}(t)$ and variance $V(t)$ of $X(t)$ are given by

$$\bar{X}(t) = E\{X(t)\} = \int x p(x, t) \, dx,$$

$$V(t) = E\{(X(t) - \bar{X}(t))^2\}. \tag{1.4.5}$$

1.4. MEAN AND VARIANCE PROPAGATION IN SDEs

The basic equation is then: What can one learn about $\bar{X}(t)$ and $V(t)$ in general? That is, one imagines that at time $t = 0$, values for $\bar{X}(t)$ and $V(t)$ are given, and one asks for future predictions about the values of $\bar{X}(t)$ and $V(t)$.

For simplicity, consider $X(t)$ to be a scalar variable. Start with the linear case in which $b(t, X) = b(t)X$ and $a(t, X)$ is constant. Then (1.4.1) becomes

$$dX = b(t)X \, dt + \sqrt{a} \, dW. \tag{1.4.6}$$

Averaging (1.4.6) by multiplying by the density and integrating gives

$$E\{dX\} = E\{b(t)X\} \, dt + E\{\sqrt{a} \, dW\}. \tag{1.4.7}$$

If $d\bar{X}$ is defined to be $E\{dX\}$, then since $\sqrt{a}$ is constant and $E\{dW\} = 0$, (1.4.7) becomes

$$d\bar{X} = b(t)\bar{X} \, dt. \tag{1.4.8}$$

Thus the mean evolves according to

$$d\bar{X}/dt = b(t)\bar{X}. \quad \text{with} \quad \bar{X}(0) = \bar{X}_0 \equiv \int x \rho_0(x) \, dx. \tag{1.4.9}$$

Equation (1.4.9) shows that the mean of $X(t)$ satisfies the same deterministic equation that is obtained by setting $a \equiv 0$ in (1.4.1). Unlike the solution of (1.4.1) with $a \equiv 0$, the variance of $X(t)$ is not zero and will change in time.

To obtain the equation for $V(t)$, we begin by solving (1.4.6). That is, rewrite (1.4.6) as

$$dX - b(t)X \, dt = \sqrt{a} \, dW, \tag{1.4.10}$$

which is equivalent to

$$\left[\frac{d}{dt}(X(t)e^{-B(t)})\right] dt = \sqrt{a} e^{-B(t)} \, dW. \tag{1.4.11}$$

In this equation

$$B(t) = \int_0^t b(s) \, ds. \tag{1.4.12}$$

Solving (1.4.11) gives

$$X(t)e^{-B(t)} - X_0 = \int_0^t \sqrt{a} e^{-B(s)} \, dW, \tag{1.4.13}$$

so that

$$X(t) = X_0 e^{B(t)} + \int_0^t e^{B(t) - B(s)} \sqrt{a} \, dW. \tag{1.4.14}$$

Since $\bar{X}(t) = X_0 e^{B(t)}$ and $V(t) = E\{[X(t) - \bar{X}(t)]^2\}$, (1.4.14) leads to

$$V(t) = E\{(X(t) - \bar{X}(t))^2\}$$
$$= E\left\{\left(\int_0^t e^{B(t) - B(s)} \sqrt{a}\, dW\right)^2\right\}. \quad (1.4.15)$$

Rewriting (1.4.15) as the product of two integrals, taking the expectations, and differentiating gives the equation

$$dV/dt = 2b(t)V(t) + a\sigma^2(t). \quad (1.4.16)$$

The initial condition for (1.4.16) is

$$V(0) = V_0 = \int (x - \bar{X}_0)^2 \rho_0(x)\, dx. \quad (1.4.17)$$

Observe that if $b(t) \equiv 0$ and $\sigma^2(t)$ are constant, then (1.4.16) shows that $V(t) = V_0 + a\sigma^2 t$. Under these assumptions, (1.4.6) becomes $dX = \sqrt{a}\, dW$, so that $X(t) = \sqrt{a}\, W(t)$, that is, $X(t)$ is simply Brownian motion times a constant. In that case one expects that the variance of $X(t)$ will be $a\sigma^2 t$.

Exercise 1.4.1

Repeat these calculations for the SDE

$$dX = b(t)X(t)\, dt + a(t)\, d\pi. \quad (1.4.18)$$

The technique that was just employed relies heavily on the linearity of the equation. For the general nonlinear equation

$$dX = b(t, X)\, dt + \sqrt{a(t, X)}\, dW, \quad (1.4.19)$$

this method will not work. However, one can try to mimic the spirit of the preceding calculations. Instead of deriving exact equations for $\bar{X}(t)$ and $V(t)$, in the nonlinear case there are approximate equations for $\bar{X}(t)$ and $V(t)$. A slight generalization of the linear case is used in the nonlinear case, so it is considered first. Let

$$dX = \beta X\, dt + \sqrt{\alpha}\, d\tilde{W}, \quad (1.4.20)$$

where $d\tilde{W}$ is a Brownian motion with a nonzero mean. In particular, let

$$E\{d\tilde{W}\} = \mu_w\, dt + o(dt),$$
$$E\{d\tilde{W}^2\} = \sigma^2\, dt + o(dt), \quad (1.4.20')$$

where the dependence of σ^2 on t is suppressed.

1.4. MEAN AND VARIANCE PROPAGATION IN SDEs

The generalizations of (1.4.9) and (1.4.16) are easily found. Observe that one can write $d\tilde{W} = \mu_w\, dt + dW$. Applying the same procedures that lead to (1.4.9) and (1.4.16) leads to

$$d\bar{X} = \beta\bar{X} + \mu_w\sqrt{\alpha},$$
$$dV/dt = 2\beta V + \alpha\sigma^2. \tag{1.4.21}$$

To obtain equations for $\bar{X}(t)$ and $V(t)$ in the nonlinear case, $b(t, X)$ and $a(t, X)$ are Taylor-expanded around $\bar{X}(t)$. The order of the Taylor expansion determines the order of the mean–variance algorithm.

The *first-order algorithm* is obtained by Taylor-expanding (1.4.19) around $\bar{X}(t)$ (which is still unknown) as follows:

$$dX = [b(t, \bar{X}) + b_x(t, \bar{X})(X - \bar{X})]\, dt + \sqrt{a(t, \bar{X})}\, dW. \tag{1.4.22}$$

Equation (1.4.22) can be made to look like (1.4.20) by writing it as

$$dX = b_x(t, \bar{X})X\, dt + d\tilde{W}, \tag{1.4.23}$$

where $d\tilde{W}$ is defined by

$$E\{d\tilde{W}\} = [b(t, \bar{X}) - b_x(t, \bar{X})\bar{X}]\, dt + o(dt),$$
$$E\{(d\tilde{W})^2\} = \sigma^2 a(t, \bar{X})\, dt + o(dt)$$

Applying (1.4.21) to (1.4.23) gives

$$d\bar{X}/dt = b(t, \bar{X}),$$
$$dV/dt = 2b_x(t, \bar{X})V + a(t, \bar{X})\sigma^2. \tag{1.4.24}$$

In the first-order algorithm, the mean evolves according to the deterministic dynamics and the variance follows a slight generalization of (1.4.17). The effects of the noise on the nonlinear dynamics are really not shown at this level. In fact, the only difference between this case and the previous one is that $b(t)$ in (1.4.9) or (1.4.17) is replaced by $b_x(t, \bar{X})$. To really observe the effects of nonlinearity, one must go to higher-order mean–variance algorithms.

In the *second-order algorithm*, (1.4.19) is Taylor-expanded using two terms. That is,

$$b(t, X) = [b(t, \bar{X}) + b_x(t, \bar{X})(X - \bar{X})$$
$$+ \tfrac{1}{2}b_{xx}(t, \bar{X})(X - \bar{X})^2] \tag{1.4.25}$$
$$+ O[(X - \bar{X})^3].$$

Using (1.4.24) in (1.4.19) and averaging gives

$$d\bar{X}/dt = b(t, \bar{X}) + \tfrac{1}{2}b_{xx}(t, \bar{X})V(t). \tag{1.4.26}$$

The effect of the noise on the nonlinear dynamics appears in (1.4.26) through the addition of a term involving the variance of the dynamics of the evolution of the mean.

The second-order algorithm for $V(t)$ is somewhat harder to derive; a nice derivation is found in the book by Sage and Melsa (1979, pp. 106–110). They show that

$$dV/dt = 2b_x(t, \bar{X})V + a(t, \bar{X})\sigma^2(t) + \tfrac{1}{2}\sigma^2(t)a_{xx}(t, \bar{X})V. \qquad (1.4.27)$$

Note how (1.4.27) differs from (1.4.24).

Example 1.4.2

For the system

$$\dot{X} = -X^3 + \xi(t), \qquad \bar{X}(0) = 0, \qquad V(0) = 0, \qquad (1.4.28)$$

the first- and second-order algorithms are, respectively,

$$d\bar{X}/dt = -\bar{X}^3, \qquad dV/dt = -6\bar{X}^2 V(t) + \sigma^2(t); \qquad (1.4.29)$$

$$d\bar{X}/dt = \bar{X}^3 - 3\bar{X}V(t), \qquad dV/dt = -6\bar{X}^2 V(t) + \sigma^2(t). \qquad (1.4.30)$$

Exercise 1.4.3

The algorithms developed in this section are approximate. Consequently, they are valid for only a limited time. One would expect the higher-order algorithm to be more accurate. To get an idea of the level of accuracy, solve (1.4.29) and (1.4.30) and compare their solutions.

Exercise 1.4.4

Find the first-and second-order algorithms for the logistic equation

$$dX = rX(1 - X/k)\,dt + \sqrt{a(t, X)}\,dW, \qquad (1.4.31)$$

where $a(t, X) = a$ (constant), aX, and aX^2.

Exercise 1.4.5

Repeat all the calculations leading to the first- and second-order algorithms for

$$dX = b(t, X)\,dt + a(t, X)\,d\pi. \qquad (1.4.32)$$

Example 1.4.6

As a warm-up for the rest of the book, now assume that the dynamics depend on a control $u(t)$. For example, in (1.4.19) replace $b(t, X)$ by

$b(t, X, u)$, where u is a control variable. Equation (1.4.19) is replaced by

$$dX = b(t, X, u)\,dt + \sqrt{a(t, X)}\,dW. \tag{1.4.33}$$

The first-order algorithm is now

$$d\bar{X}/dt = b(t, \bar{X}, u), \qquad dV/dt = 2b_x(t, \bar{X}, u)V + a(t, \bar{X})\sigma^2. \tag{1.4.34}$$

One can conceive of many control problems for this system, although not as many can be solved.

For example, in portfolio management problems one often deals with mean–variance models. These models, in the context of (1.4.33), take the form

$$\min \operatorname{Var}\{X(T)\} \quad \text{such that} \quad E\{X(T)\} \geq x. \tag{1.4.35}$$

Here T is a given end time and x a given target level for $X(T)$. This *stochastic control* problem is converted to a deterministic problem by (1.4.34). The deterministic control problem is

$$\min_{u} V(T) \quad \text{such that (1.4.34) holds and} \quad \bar{X}(T) \geq x. \tag{1.4.36}$$

Problem (1.4.36) can be solved using optimal control theory, which is reviewed in Chapter 3.

Exercise 1.4.7

Apply optimal control theory to obtain a solution of (1.4.36).

Bibliographic Notes

The books by Sage and Melsa (1979) and Bryson and Ho (1975) contain nice treatments of the algorithms described in this section. The use of such approximate methods is common in the engineering literature.

1.5 KALMAN FILTERING AND ITS EXTENSIONS

In general, the mean–variance algorithms (1.4.24), (1.4.26) and (1.4.27) derived in the previous section are difficult to analyze. One property is clear, however. In both cases $a(t, \bar{X})\sigma^2(t)$ forces the equation for $\bar{V}(t)$. This means that the variance $V(t)$ increases with t. One way of reducing the variance is to make measurements on the system. When the measurements themselves contain noise, the Bayes theorem can be used to update the combination of system dynamics and measurement. This general process is called *filtering*. If the state dynamics are treated in the time domain, it is known as *Kalman filtering*. If the state dynamics are treated in the frequency domain, it is known as *Wiener filtering*. Kalman's original paper (Kalman, 1960) still contains one of the best descriptions of the problems and viewpoints.

Suppose that $X(t)$ satisfies the stochastic differential equation

$$dX = b(t, X)\, dt + \sqrt{a(t, X)}\, dW. \tag{1.5.1}$$

Assume that $X(0)$ is not known exactly but that $X(0)$ itself is normally distributed with mean x_0 and variance σ_0^2. The goal of making measurements is to reduce the uncertainty about $X(t)$.

To begin, assume that $X(t)$ is measured at a discrete set of points $\{t_k\}$, $k = 1, 2, \ldots$. The measurement obtained at t_k, denoted Z_k, is a function of t_k, $X(t_k)$, and a noise term

$$Z_k = h(t_k, X(t_k)) + v_k. \tag{1.5.2}$$

Here $h(s, y)$ is a known function and $\{v_k\}$ a sequence of independent normal random variables with mean 0 and variance R_k. It is assumed that v_k and $W(t)$ are independent. The measurement function $h(s, y)$ may be linear, but often it is not. Many applications are dominated by trying to force truly nonlinear problems into a linear form, which is not the best approach. The next example shows how easily nonlinear problems arise.

Example 1.5.1

In Chapter 5, sighting surveys for whale and porpoise populations are discussed. In such cases, an observer records two bearings and a distance traveled (see Fig. 1.5.1) and uses them to estimate a range. (The problem is actually more complicated than this, and the operation slightly different, but this model is sufficient as an illustration.) From Fig. 1.5.1,

$$R = (\sin \theta_1) S / \sin(\theta_2 - \theta_1). \tag{1.5.3}$$

When such a sighting survey is actually performed, it is usually easy to measure S (which is the distance traveled), since the speed of the vessel and the time between observations are easily measured. On the other hand, an accurate measurement of θ_i is more difficult. Typically, the measurements of θ_i are corrupted by noise. In actuality, R is not used, but R_L, the lateral range from the observer to the whales, is needed. Since $R_L = R \sin \theta_2$, the final measurement relation is

$$R_L = [\sin \theta_1 \sin \theta_2 / \sin(\theta_2 - \theta_1)] S, \tag{1.5.4}$$

1.5. KALMAN FILTERING AND ITS EXTENSIONS

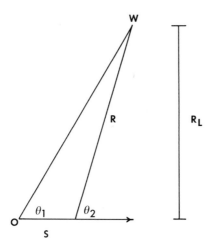

Fig. 1.5.1 The whale-sighting problem. An observer O sights a whale W at angle θ_1 and then travels a distance S and observes the whale at angle θ_2. The distance between the whale and observer after the second sighting is R; the lateral (perpendicular) range is R_L.

which is highly nonlinear. One must deal with this nonlinear measurement function, since there is no way to avoid it.

The question is then, how should the measurement information be incorporated? As before, it pays to tackle the linear case first. The linear case corresponds to

$$b(t, X) = b(t)X, \quad a(t, X) = a \text{ (const.)} \quad h(t, X) = h(t)X. \quad (1.5.5)$$

The mean $\bar{X}(t)$ evolves according to

$$d\bar{X}/dt = b(t)\bar{X}. \quad (1.5.6)$$

Let $\bar{X}_k(-)$ be the value of $\bar{X}(t_k)$ before a measurement is taken. At that time, the estimate of the variance is given by the solution of (1.4.16). Let $V_k(-)$ be the value of $V(t_k)$ before the measurement is taken. At time t_k an observation Z_k is taken and an *update* of $\bar{X}(t_k)$ is computed. Let $\bar{X}_k(+)$ be the value of $\bar{X}(t_k)$ after the measurement is taken. Assume that $\bar{X}_k(+)$ takes the form

$$\bar{X}_k(+) = a_k + K_k Z_k. \quad (1.5.7)$$

In this equation, a_k and K_k are to be determined. Set

$$\tilde{X}_k(\pm) = \bar{X}_k(\pm) - X(t_k). \quad (1.5.8)$$

Substituting (1.5.7) in (1.5.8) gives

$$\tilde{X}_k(+) = a_k + K_k Z_k - X(t_k)$$
$$= a_k + K_k Z_k + \tilde{X}_k(-) - \bar{X}_k(-). \tag{1.5.9}$$

Observe from (1.5.8) that $\tilde{X}_k(\pm)$ is the deviation between the mean and the true value of $X(t)$ and thus is a random variable. On the average, one wants it to be unbiased, that is, $E\{\tilde{X}_k(\pm)\} = 0$. Since $Z_k = h_k X(t_k)$ and $E\{v_k\} = 0$, averaging (1.5.9) gives

$$E\{\tilde{X}_k(+)\} = a_k + K_k h_k \bar{X}_k(-) - \bar{X}_k(-). \tag{1.5.10}$$

Setting the right-hand side of (1.5.10) equal to 0 and solving for a_k gives

$$a_k = \bar{X}_k(-) - K_k h_k \bar{X}_k(-). \tag{1.5.11}$$

Substituting (1.5.11) in (1.5.7) shows that the unbiased estimate for the update is (K_k is still undetermined)

$$\bar{X}_k(+) = \bar{X}_k(-) + K_k\{Z_k - h_k \bar{X}_k(-)\}. \tag{1.5.12}$$

Note that if K_k or h_k were zero, then $\bar{X}_k(+) = \bar{X}_k(-)$, as expected. In this case, no measurement is taken and no information is obtained.

How should K_k be picked? One choice is to pick K_k so that the variance of $\tilde{X}_k(+)$ is a minimum. Then (1.5.7) is the minimum-variance unbiased estimate of $\tilde{X}(t_k^+)$. Setting

$$\partial/\partial K_k \, \text{Var}\{\tilde{X}_k(+)\} = 0 \tag{1.5.13}$$

and solving for K_k gives

$$K_k = V_k(-) h_k \{h_k^2 V_k(-) + R_k\}^{-1}. \tag{1.5.14}$$

In order to obtain (1.5.14), the distribution on v_k (normal with mean 0 and variance R_k) has to be used. Equation (1.5.14) provides a way to meld system dynamics and measurement information.

The procedure just described is called the *linear Kalman filter*. If t_k^- denotes the time just before the measurement and t_k^+ the time just after the measurement, the algorithm is summarized as follows.

The state estimate is

$$d\bar{X}/dt = b(t)\bar{X}, \quad \bar{X}(t_k^+) = \bar{X}_k(+), \quad t_k \leq t < t_{k+1}. \tag{1.5.15}$$

The variance estimate is

$$dV/dt = 2b(t)V + a\sigma^2(t), \quad V(t_k^+) = V_k(+), \quad t_k \leq t < t_{k+1}. \tag{1.5.16}$$

1.5. KALMAN FILTERING AND ITS EXTENSIONS

The updating algorithm is

$$\bar{X}(t_k^+) \equiv \bar{X}_k(+) = \bar{X}_k(-) + K_k\{Z_k - h_k\bar{X}_k(-)\},$$
$$V(t_k^+) \equiv V_k(+) = [1 - K_k h_k]V_k(-), \quad (1.5.17)$$
$$K_k = V_k(-)h_k[h_k^2 V_k(-) + R_k]^{-1}.$$

Exercise 1.5.2

Describe what happens if (1.5.1) is replaced by

$$dX = b(t)X\,dt + a(t)\,d\pi. \quad (1.5.18)$$

Exercise 1.5.3

Describe what happens if (1.5.7) is replaced by

$$\bar{X}_k(+) = a_k + b_k Z_k + c_k Z_k^2. \quad (1.5.19)$$

Instead of taking measurements at discrete times, the measurements could be taken continuously. In this case (1.5.2) is replaced by

$$Z(t) = h(t)X(t) + v(t). \quad (1.5.20)$$

Here $v(t) \sim N(0, R(t))$, and $v(t)$ is assumed to be independent of $w(t)$. The simplest way to treat the continuous time case is as a limit, namely, set $t_{k+1} = t_k + \Delta t$. Then define

$$\Delta \bar{X}(t_{k+1}) = \bar{X}(t_{k+1}^+) - \bar{X}(t_k^+),$$
$$\Delta V(t_{k+1}) = V(t_{k+1}^-) - V(t_k^-). \quad (1.5.21)$$

Taking Eqs. (1.5.15)–(1.5.21), dividing by Δt, and letting $\Delta t \to 0$ gives the state estimate

$$d\bar{X}/dt = b(t)\bar{X}(t) + K(t)\{Z(t) - h(t)\bar{X}(t)\}, \quad (1.5.22)$$

the variance propagation

$$dV/dt = 2b(t)V + a\sigma^2(t) - \tfrac{1}{2}K(t)^2 R(t), \quad (1.5.23)$$

and the Kalman gain

$$K(t) = V(t)h(t)/R(t). \quad (1.5.24)$$

One advantage of Kalman filtering in continuous time is that (1.5.23) can be solved independently of (1.5.22); the variance and gain can be computed without using the data.

Exercise 1.5.4

The filtering algorithms that were just derived were based on the assumption that the state noise $W(t)$ and measurement noise v_k (in the discrete case) or $v(t)$ (in the continuous case) were independent. Rederive the filtering algorithms for the case in which $W(t)$ and the measurement noise are not independent.

Exercise 1.5.5

The Ornstéin–Uhlenbeck process satisfies the equation

$$dX = -\alpha X \, dt + dW. \tag{1.5.25}$$

In this case, it is actually possible to find the density for $X(t)$ directly [see Feller (1971, pp. 335–336)]. If $X(0) = x_0$ and $W(t)$ is normalized so that $E\{(dW)^2\} = dt$, then $X(t)$ is distributed according to

$$X(t) \sim N(x_0 e^{-\alpha t}, (1 - e^{-2\alpha t})/2\alpha). \tag{1.5.26}$$

Note that the limiting distribution, as $t \to \infty$, is $N(0, \tfrac{1}{2}\alpha)$.

Suppose that x_0 is a random variable and that measurements are taken according to the formula

$$Z = X \, dt + g \, dv. \tag{1.5.27}$$

Here $v(t)$ is another Brownian motion process. If $E\{x_0\} = \bar{x}_0$ and $\text{Var}\{x_0\} = \bar{v}_0$, the equations for the filter are

$$d\bar{X} = -\alpha \bar{X} \, dt + [V(t)/g^2](Z - \bar{X} \, dt), \qquad \bar{X}(0) = \bar{x}_0, \tag{1.5.28}$$

and

$$dV/dt = 1 - V^2/g^2 - 2V\alpha, \qquad V(0) = \bar{v}_0. \tag{1.5.29}$$

The equation for the variance can be solved to give

$$V(t) = V_\infty + [(V_\infty - V_1)(\bar{v}_0 - V_\infty)]/[(\bar{v}_0 - V_1)e^{2\alpha t} - (\bar{v}_0 - V_\infty)], \tag{1.5.30}$$

where

$$\beta = (\alpha + 1/g^2)^{1/2}, \qquad V_\infty = g^2(\beta - \alpha), \qquad V_1 = g^2(\beta + \alpha). \tag{1.5.31}$$

As $t \to \infty$, $V(t) \to V_\infty$, which is a measure of the limiting precision for observations in the system.

Example 1.5.5

Often the precision of a measurement is related to the cost. This example shows how the cost and precision can be balanced in an optimal fashion. Suppose that the state dynamics and measurement equation are given by

$$dX/dt = f(t)X(t), \qquad X(0) \sim N(x_0, V_0). \tag{1.5.32}$$

1.5. KALMAN FILTERING AND ITS EXTENSIONS

According to (1.5.32), $X(t)$ evolves deterministically from an unknown initial value. Measurements on $X(t)$ could be used to help determine the actual value. In particular, assume that the measurement function is given by

$$Z(t) = h(t)X(t) + v(t). \tag{1.5.33}$$

Assume that $v(t) = N(0, R(t))$ and that the precision of the measurement depends on a control $u(t)$ in the following way:

$$h(t) = u(t)^{n/2}, \qquad 1/R(t) = \beta u(t)^m / [\gamma + u(t)^m]. \tag{1.5.34}$$

Here n, m, β, and γ are fixed constants.

The variance $V(t)$ for the filter satisfies

$$dV/dt = 2f(t)V(t) - V(t)^2 h(t)^2 / R(t). \tag{1.5.35}$$

Now consider a time period $(0, T)$. Since $X(0)$ is not known precisely, $X(T)$ will not be known precisely. Assume, however, that one wishes to know the value of $X(T)$ as accurately as possible, at minimum cost. This suggests forming a cost functional

$$J = c_1 V(T) + c_2 \int_0^T u(s)\, ds. \tag{1.5.36}$$

Here c_1 and c_2 are the costs of imperfect knowledge of the state and of applying the control, respectively. The problem of interest is then

$$\begin{aligned}&\text{minimize} \quad c_1 V(T) + c_2 \int_0^T u(s)\, ds \quad \text{such that} \\ &dV/dt = 2f(t)V(t) - V(t)^2 \beta u(t)^{n+m}/[\gamma + u(t)^m].\end{aligned} \tag{1.5.37}$$

The solution of this example can be found by applying optimal control theory. It is discussed more fully in Chapter 3, after optimal control theory is treated.

When the system dynamics are nonlinear or the state measurement is nonlinear, one of the mean–variance algorithms of the previous section must be used. In addition, the measurement must be treated in a similar fashion. For example, if the measurement function is

$$Z_k = h(t_k, X(t_k)) + v_k, \tag{1.5.38}$$

then the reasoning associated with the first-order mean–variance algorithm suggests an expansion of the form

$$\begin{aligned}Z_k = {}&h(t_k, \bar{X}(t_k^-)) + h_x(t_k, \bar{X}(t_k^-))(X(t_k) - \bar{X}(t_k^-)) \\ &+ O((X(t_k) - \bar{X}(t_k^-))^2).\end{aligned} \tag{1.5.39}$$

Using (1.5.38) leads to an algorithm of the following form:
The state estimate is

$$d\bar{X}/dt = b(t, \bar{X}(t)), \qquad t_k \leq t < t_{k+1}; \qquad (1.5.40)$$

the variance propagation is

$$dV/dt = 2b_x(t, \bar{X}(t))V + a\sigma^2(t), \qquad t_k \leq t < t_{k+1}; \qquad (1.5.41)$$

the updating algorithm is

$$\bar{X}(t_k^+) = \bar{X}(t_k^-) + K_k\{Z_k - h(t_k, \bar{X}(t_k^-))\},$$
$$V(t_k^+) = (1 - K_x h_x(t_x, \bar{X}(t_k^-)))V(t_k^-); \qquad (1.5.42)$$

and the Kalman gain is

$$K_k = V(t_k^-)h_x(t_k, \bar{X}(t_k^-))\{h_x(t_k, \bar{X}(t_k^-))^2 V(t_k^-) + R_k\}^{-1}.$$

For measurements that are taken in continuous time, the algorithm is somewhat simpler: The state estimate is

$$d\bar{X}/dt = b(t, \bar{X}) + K(t)\{Z(t) - h(t, \bar{X}(t))\}, \qquad (1.5.43)$$

the variance propagation is

$$dV/dt = 2b_x(t, \bar{X})V + a\sigma^2(t) - [V(t)^2 h_x(t, \bar{X})^2/R(t)], \qquad (1.5.44)$$

and the Kalman gain is

$$K(t) = V(t)h_x(t, \bar{X}(t))/R(t). \qquad (1.5.45)$$

Example 1.5.6

The following equations are a model for forest growth (Clark, 1976):

$$dX/dt = at^{-b}Xe^{-cX}, \qquad X(t_0) = X_0. \qquad (1.5.46)$$

Here $X(t)$ is the volume of timber of an age greater than or equal to t_0 and a, b, and c are constants. To include stochastic effects, (1.5.46) can be replaced by

$$dX = at^{-b}Xe^{-cX}\,dt + \sqrt{\varepsilon(t, x)}\,dW, \qquad (1.5.47)$$

where $\varepsilon(t, x)$ is a known function. Assume that measurements are taken according to

$$Z(t_k) = h(t_k)X(t_k) + v_k. \qquad (1.5.48)$$

If $c = 0$, then the linear filtering equations apply. Otherwise, the nonlinear algorithms must be used.

1.5. KALMAN FILTERING AND ITS EXTENSIONS

Exercise 1.5.7

Solve and compare the filter equations in the linear ($c \equiv 0$) and linearized (that is, when Xe^{-cX} is Taylor-expanded) cases.

Bibliographic Notes

The books by Gelb (1974), Bryson and Ho (1975), and Sage and Melsa (1979) contain good chapters on filtering. The first- and second-order mean–variance algorithms and their use in filtering are described by Gelb (1974) and Sage and Melsa (1979). Example 1.5.5 is similar to a paper by Athans (1972). The paper by Uhlenbeck and Ornstein (1930), which is reprinted in Wax (1954), is a true gem of analysis. It is worth understanding the Uhlenbeck–Ornstein process, because it has so many different applications. Example (1.5.4) (on filtering for the Uhlenbeck–Ornstein process) comes from Davis (1977, pp. 141–142). The papers of Dixon and Howitt (1979a,b) contain an excellent application of filtering theory to forest management.

2

Stochastic Control and Dynamic Programming

The necessary background in probability and stochastic processes has now been developed. In many of the problems considered in this book, the objective is to maximize a functional of one or more stochastic variables. In general, such problems are solved using stochastic dynamic programming.

2.1 THE PRINCIPLE OF OPTIMALITY

Much of the work on stochastic control is based on the *principle of optimality*. A rough statement of the principle is the following. Suppose that $u^*(s)$, $\tau \leq s \leq T$ is the optimal control for a problem, starting at $s = \tau$, with $X(\tau)$ given. Consider a subproblem, starting at $s = \tau_1 > \tau$. Then $u^*(s)$, $\tau_1 \leq s \leq T$, is also optimal for this problem.

Figure 2.1.1 illustrates this concept. As stated, this principle does not help the analyst, who wants to find the optimal control starting at times $\tau_0 < \tau$. What the principle of optimality does do, however, is provide a method for deriving an equation (the dynamic programming equation) that can be used to find the optimal control. To get an idea of how such equations are derived, suppose that one has the deterministic problem

$$\max_u \left\{ \sum_{j=T}^{M} \alpha(j) v(j, u) \right\}. \tag{2.1.1}$$

2.1. THE PRINCIPLE OF OPTIMALITY

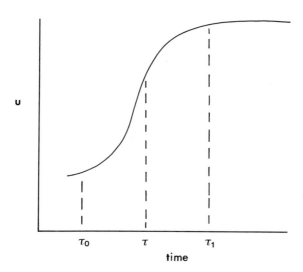

Fig. 2.1.1 An Illustration of the principle of optimality. The function $u(t)$ is the optimal control as a function of time. If $u(t)$ is optimal starting at τ, then it is clearly also optimal starting at τ_1.

Here $\alpha(j)$ is a known weighting function, u a control, and $v(j, u)$ the "value function" at time j. Define a function $V(T, u)$ by

$$V(T, u) = \sum_{j=T}^{M} \alpha(j)v(j, u); \qquad (2.1.2)$$

observe that problem (2.1.1) is then the same as

$$\max_{u}\{\alpha(T)v(T, u) + V(T + 1, u)\}. \qquad (2.1.3)$$

In particular, the two can be combined to give

$$\max_{u} V(T, u) = \max_{u}\{\alpha(T)v(T, u) + V(T + 1, u)\}. \qquad (2.1.4)$$

Equation (2.1.4) is solved recursively. It should be noted that the recursion actually goes "backwards" in time, since one obtains $V(T, u)$ in terms of $V(T + 1, u)$. Hence, this process is sometimes called "backwards induction." Equation (2.1.4), which contains the principle of optimality, is a simple deterministic dynamic programming equation. In this chapter similar dynamic programming equations are derived for various kinds of dynamic and stochastic assumptions.

In general, the control problems have these components:

(1) stochastic dynamics for the state $X(t)$,
(2) a value function depending on the state $X(t)$ and control $u(t)$,
(3) a terminal value function depending on $X(T)$.

The goal of the theory is to find the control that optimizes the value function, subject to the appropriate dynamics.

It is useful at this point to make a distinction between certain dynamics and uncertain dynamics. In the first case, one assumes that all parameters of the process of interest are known, but that the system is subject to random fluctuations. In the second case, some of the parameters themselves are unknown and the dynamics may also be subject to random fluctuations. It turns out that the case in which all parameters are known but the dynamics involve fluctuations is the easier case to deal with, so it is a good starting point.

Bibliographic Notes

There are many books on dynamic programming and stochastic control. The books by Fleming and Rishel (1975) and Bertsekas (1976) provide good introductions. In particular, the book by Fleming and Rishel is slightly more abstract and mathematical than that by Bertsekas. Kleindorfer (1978) provides a review and general history of dynamic programming, as well as examples in management science. Sengupta (1982) provides other examples in resource management and economics. Ross (1983) gives a moderate-level mathematical treatment.

2.2 THE DYNAMIC PROGRAMMING EQUATION FOR A $W(t)$-DRIVEN PROCESS

We begin by studying the dynamic programming equation (DPE) when the state equation contains Brownian motion. The results of this section are used in Section 2.4 on finding optimal controls directly, 2.5 on control with probability criteria, 4.4 on optimal exploration for an exploitation of uncertain exhaustible resource markets, 4.5 on the price dynamics of exhaustible resources, 6.2 on harvesting a randomly fluctuating population, 6.4 on the trade-off between stock and control fluctuations, and 6.5 on the price dynamics of renewable resource markets.

Assume that the state $X(t)$ satisfies the stochastic differential equation

$$dX = b(t, X, u)\, dt + a(t, X, u)\, dW. \qquad (2.2.1)$$

2.2. THE DPE FOR A W(t)-DRIVEN PROCESS

Here $u(t)$ is a control in some specified control set V. For simplicity, assume that $E\{(dW)^2\} = dt$. Assume that the objective functional is

$$J(x_0) = \max_{u \in V} E\left\{\int_0^T e^{-rs}L(s, X, u)\, ds + \varphi(X(T))\,|\,X(0) = x_0\right\}. \quad (2.2.2)$$

In (2.2.2) it is understood that $X = X(s)$ and that r, $L(\cdot, \cdot, \cdot)$, and $\varphi(\cdot)$ are given. For example, $X(t)$ could be the stock of resource at time t, $u(t)$ the rate of exploitation of the resource at time t, $L(s, X, u)$ the economic rent or social utility obtained through exploitation of the resource, $\varphi(\cdot)$ a preservation value of the resource, and r the discount rate.

To find $J(x_0)$, we introduce a new function defined by

$$J(x, t) = \max_{u \in V} E\left\{\int_t^T e^{-rs}L(s, X, u)\, ds + \varphi(X(T))\,|\,X(t) = x\right\}. \quad (2.2.2')$$

Clearly $J(x, T) = \varphi(x)$; the principle of optimality can be used to derive a differential equation for $J(x, t)$. To do this, use the principle with $\tau = t$ and $\tau_1 = t + dt$. Then, breaking up the integral in (2.2.2') gives

$$J(x, t) = \max_{u \in V} E\left\{\int_t^{t+dt} e^{-rs}L(s, X, u)\, ds\right.$$

$$\left.+ \int_{t+dt}^T e^{-rs}L(s, X, u)\, ds + \varphi(X(T))\,|\,X(t) = x\right\}. \quad (2.2.3)$$

In $(t, t + dt)$ the process moves from x to $x + dX$, where the random increment dX is given by (2.2.1). The second integral in (2.3) is $J(x + dX, t + dt)$, so that $J(x, t)$ can also be written as

$$J(x, t) = \max_{u \in V} E_{dX}\left\{\int_t^{t+dt} e^{-rs}L(s, X, u)\, ds + J(x + dX, t + dt)\,|\,X(t) = x\right\}. \quad (2.2.4)$$

A Taylor expansion of (2.2.4) around (x, t) using $E\{o(dX^2)\} = o(dt)$ gives

$$J(x, t) = \max_{u \in V} E_{dX}\{e^{-rt}L(t, x, u)\, dt + o(dt)$$

$$+ J(x, t) + J_t\, dt + J_x\, dx + \tfrac{1}{2}J_{xx}(dX)^2 + o(dt)\}. \quad (2.2.5)$$

Taking the expectation over dX, diving by dt, and letting $dt \to 0$ gives the partial differential equation

$$0 = J_t + \max_{u \in V}\{e^{-rt}L(t, x, u) + b(t, x, u)J_x + \tfrac{1}{2}a(t, x, u)J_{xx}\}. \quad (2.2.6)$$

Equation (2.2.6) is the desired dynamic programming equation. It satisfies the end condition

$$J(x, T) = \varphi(x). \qquad (2.2.7)$$

In order to obtain (2.2.7), observe that when $t = T$ in (2.2.2') the integral vanishes and all that remains is $\varphi(X(T))$. Since $X(T) = x$, one immediately obtains (2.2.7). The discussion of boundary conditions is deferred.

Sometimes it is possible to simplify the DPE. For example, if $L(\cdot, \cdot, \cdot)$, $b(\cdot)$, and $a(\cdot)$ are functions of x and u only, one eliminates the time dependence in (2.2.6) by setting $J(x, t) = W(x)e^{-rt}$. Then $J_t = -rW(x)e^{-rt}$ and the DPE becomes

$$0 = -rW(x) + \max_{u \in V}\{L(x, u) + b(x, u)W_x + \tfrac{1}{2}a(x, u)W_{xx}\}. \qquad (2.2.8)$$

Two technical points are in order. First, to reach the DPE (2.2.6) or (2.2.8), it was assumed that $J(x, t)$ exists and is twice differentiable. The existence and differentiability have not been demonstrated, and they present thorny analytical problems. Second, (2.2.6) and (2.2.8) usually are nonlinear equations. Hence, solutions need not be unique and may not exist for all x and t. The message of these points is simply to proceed with caution.

Exercise 2.2.1

Repeat all these calculations in *discrete time*, namely, assume that

$$X(k + 1) = b(k, X(k), u(k)) + W_k, \qquad (2.2.9)$$

where $u(k)$ is a control and W_k a noise term with

$$W_k \sim N(0, a(k, X(k), u(k))).$$

Assume that the objective functional is

$$J(x, l) = E\left\{\sum_{k=l}^{K} L(k, X(k), u(k))(1 + r)^{-k} + \varphi(X(K)) \mid X(l) = x\right\}. \qquad (2.2.10)$$

Derive a difference equation for $J(x, l)$.

It is rare indeed that an analytic solution of the DPE can be found; two examples in which solutions are found follow.

Example 2.2.2: The Linear Regulator

One completely solvable case is the "linear regulator." The state equation is linear in the state variable and the control:

$$dX = [A(t)X(t) + B(t)u(t)]\, dt + \sqrt{\sigma(t)}\, dW. \qquad (2.2.11)$$

2.2. THE DPE FOR A $W(t)$-DRIVEN PROCESS

The objective function (often called a performance index) is

$$J(x, t) = \min_{u \in V} E \left\{ \int_t^T [M(t')X(t')^2 + N(t')u(t')^2]\, dt' + DX(T)^2 \,|\, X(t) = x \right\}. \tag{2.2.12}$$

The form of the objective functional shows that the object of control in this example is to get the process near the origin and keep it there with a minimum use of the control. In this case, the DPE (2.2.6) becomes

$$0 = J_t + \min_{u \in V} \{M(t)x^2 + N(t)u^2 + J_x(A(t)x + B(t)u) + \tfrac{1}{2}\sigma(t)J_{xx}\}. \tag{2.2.13}$$

Assuming that the minimum occurs in the interior of V, elementary calculus can be applied to (2.2.13). Then the minimizing u satisfies

$$u^*(t) = -J_x B(t)/2N(t). \tag{2.2.14}$$

Substituting (2.2.14) into (2.2.13) and simplifying gives the nonlinear equation

$$0 = J_t + M(t)x^2 + J_x(A(t)x) - B^2(t)J_x^2/4N(t) + \tfrac{1}{2}\sigma(t)J_{xx}. \tag{2.2.15}$$

The DPE (2.2.13) was linear, but involved an optimization step. Equation (2.2.15) is nonlinear, but contains no optimization step. This equation is pretty formidable. Observe, however, that the coefficients involve powers of x^0, x^1, and x^2 only. Also, observe that the powers of x and derivatives of $J(x, t)$ appear in a unique way. In particular, if $J(x, t)$ were quadratic in x, then only powers of x^0 and x^2 would appear in the equation. These observations suggest that one should seek a solution of (2.2.15) in the form

$$J(x, t) = x^2 K(t) + q(t), \tag{2.2.16}$$

with $K(t)$ and $q(t)$ to be determined.

If $J(x, t)$ is given (2.2.16), then the various derivatives are

$$J_t = x^2 K'(t) + q'(t), \qquad J_x = 2xK(t), \qquad J_{xx} = 2K(t). \tag{2.2.17}$$

Substituting (2.2.16) into (2.2.15) gives

$$0 = x^2 K'(t) + q'(t) + M(t)x^2 - B(t)^2 x^2 K^2/N(t) + 2A(t)x^2 K + \sigma(t)K. \tag{2.2.18}$$

Setting coefficients of powers of x equal to zero gives equations for $K(t)$ and $q(t)$. These equations are

$$\begin{aligned} O(x^0), &\quad q' + \sigma(t)K(t) = 0; \\ O(x^2), &\quad K'(t) + 2AK - B^2K^2/N = -M. \end{aligned} \tag{2.2.19}$$

End conditions are needed for these equations. From (2.2.12), $J(x, T) = D(T)x^2$, so that the end conditions are

$$K(T) = D(T), \quad q(T) = 0. \tag{2.2.20}$$

Exercise 2.2.3

Solve equations (2.2.19). [Hint: the equation for $K(t)$ is a Ricatti equation. If $M \equiv 0$, set $f(t) = 1/K(t)$. Otherwise, try a power series solution.]

The method described here, that is, guessing the form (2.2.16), works because of the special linear–quadratic structure of the problem. In general, such special structure is missing and the solution is harder to obtain.

Example 2.2.4: Portfolio Selection

Merton (1971) studies the following problem of optimal portfolio selection. There are two assets with prices p_1, p_2 that change according to

$$dp_1 = p_1 r \, dt, \quad dp_2 = p_2(\alpha \, dt + \sigma \, dW). \tag{2.2.22}$$

The first asset is riskless; its price grows exponentially with growth parameter r. The second asset is risky. From (2.2.22), given that $p_2(t) = \bar{p}_2$, the increment dp_2 is normally distributed with mean $\bar{p}_2 \alpha \, dt$ and variance $\bar{p}_2^2 \sigma^2 \, dt + o(dt)$ [here $W(t)$ is Brownian motion with $E\{dW^2\} = dt$].

Exercise 2.2.5

Observe that p_2 is a function $p_2(t, W)$ of t and $W(t)$ and that $d \ln p_2 = dp_2/p_2 = \alpha \, dt + \sigma \, dW$. However, $p_2(t, W)$ is not $\exp\{\alpha t + \sigma W\}$, since the Ito calculus [Eq. (1.2.5)] must be used. What is the explicit form of $p_2(t, W)$?

The question faced by an investor is: How should money be allocated between the two assets? To answer this question, let $X(t)$ be the wealth at time t, $u_1(t)$ the fraction of wealth in the risky asset, and $u_2(t)$ the consumption rate. Common sense requires that $0 \leq u_1 \leq 1$ and $u_2 \geq 0$.

The increment of wealth at time t satisfies the SDE

$$dX = (1 - u_1)Xr \, dt + u_1 X(\alpha \, dt + \sigma \, dW) - u_2 \, dt. \tag{2.2.23}$$

The performance index in this case is assumed to be given by

$$J(x, t) = \max_{u_1, u_2} E\left\{ \int_t^T e^{-\rho t} F[u_2] \, dt \,\Big|\, X(t) = x \right\}. \tag{2.2.24}$$

Here ρ is the discount factor and $F[\cdot]$ a utility function associated with the consumption of wealth.

2.2. THE DPE FOR A $W(t)$-DRIVEN PROCESS

In this case the DPE (2.2.6) becomes

$$0 = J_t + \max_{u_1, u_2}\{\tfrac{1}{2}(u_1 x\sigma)^2 J_{xx} + \{(1 - u_1)xr + u_1 x\alpha - u_2\}J_x + e^{-\rho t}F(u_2)\}. \tag{2.2.25}$$

Assuming internal controls (i.e., an internal maximum) and applying elementary calculus gives u_1^* explicitly and an implicit equation for u_2^* (here u_1^* and u_2^* are the optimal values of u_1 and u_2). These equations are

$$u_1^* = (r - \alpha)J_x/\sigma^2 x J_{xx}, \qquad F'(u_2)|_{u_2^*} = e^{\rho t} J_x. \tag{2.2.26}$$

To go any further, an explicit assumption about the form of $F(v)$ is needed. Assume that $F(v) = v^\gamma$, where $0 < \gamma < 1$ is a given parameter. This form for $F(v)$ suggests seeking a solution for (2.2.25) in the form

$$J(x, t) = g(t)x^\gamma. \tag{2.2.27}$$

[The remark in Section 2.1 about "seeking" the nature of the solution also applies to (2.2.27).] Using (2.2.27) in (2.2.26) gives u_1^* and u_2^* explicitly:

$$u_1^* = (\alpha - r)/\sigma^2(1 - \gamma), \qquad u_2^* = x\{e^{\rho t}g(t)\}^{1/(\gamma - 1)}. \tag{2.2.28}$$

Substituting (2.2.27) in (2.2.25) after using (2.2.28) gives the following equation for $g(t)$:

$$0 = \frac{dg}{dt} + \left\{\frac{(\alpha - r)^2}{2\sigma^2(1 - \gamma)} + r\right\}\gamma g + (1 - \gamma)g\{e^{\rho t}g\}^{1/(\gamma - 1)}. \tag{2.2.29}$$

In light of (2.2.24), $g(T) = 0$.

Exercise 2.2.6

To solve (2.2.29), set

$$h(t) = [e^{\rho t}g(t)]^{1/(1 - \gamma)}. \tag{2.2.30}$$

Then obtain a linear equation for $h(t)$, and solve it.

Bibliographic Notes

Fleming and Rishel (1975) contains an excellent discussion of DPEs for processes driven by Brownian motion. Karlin and Taylor (1981) discuss an example in option pricing similar to Example 2.2.4. The use of "Brownian motion" models in finance is now quite common. Smith (1976) provides a nice review of these models. Chernoff and Petkau (1978) describe another problem involving optimal control of Brownian motion.

2.3 DYNAMIC PROGRAMMING EQUATION FOR A $\pi(t)$-DRIVEN PROCESS

The Brownian motion process $W(t)$ corresponds to a continuum of changes and its DPE is a second-order partial *differential* equation. In this section the jump process $\pi(t)$ is considered; recall that $\pi(t)$ is characterized by discrete changes. The DPE turns out to be a *difference* equation.

The results in this section are used in Section 4.3 on the utilization of an uncertain resource with learning, 4.4 on optimal exploration for and exploitation of an exhaustible resource, and 5.3 on surveys of fish stocks.

Assume that the state equation is

$$dX = b(t, X, u)\,dt + a(t, X, u)\,d\pi(t). \tag{2.3.1}$$

For the jump process $d\pi(t)$, make the following assumptions:

$$\begin{aligned}\Pr\{d\pi = \delta_1\} &= \lambda(x, u)\,dt + o(dt),\\ \Pr\{d\pi = -\delta_2\} &= \mu(x, u)\,dt + o(dt),\\ \Pr\{d\pi = 0\} &= 1 - (\lambda(x, u) + \mu(x, u)))\,dt + o(dt).\end{aligned} \tag{2.3.2}$$

Here δ_1 and δ_2 are assumed to be positive, so that the three cases in (2.3.2) correspond to a jump up, a jump down, and no jump. These three events are assumed to exhaust all of the possibilities.

As before, take the objective functional to be

$$J(x, t) = \max_u \mathrm{E}\left\{\int_t^T e^{-rs}L(s, X, u)\,ds + \varphi(X(T))\,|\,X(t) = x\right\}. \tag{2.3.3}$$

Proceeding as before, the principle of optimality is applied and the integral is broken into two pieces. This gives

$$\begin{aligned}J(x, t) = \max_u \mathrm{E}\bigg\{&\int_t^{t+dt} e^{-rs}L(s, X, u)\,ds \\ &+ \int_{t+dt}^T e^{-rs}L(s, X, u)\,ds + \varphi(X(T))\,|\,X(t) = x\bigg\}.\end{aligned} \tag{2.3.4}$$

Since the second integral gives $J(x + dX, t + dt)$, (2.3.4) can be rewritten as

$$J(x, t) = \max_u \{e^{-rt}L(t, x, u)\,dt + o(dt) + \mathrm{E}_{dX}[J(x + dX, t + dt)]\}. \tag{2.3.5}$$

In Section 2.2, $J(x + dX, t + dt)$ was Taylor expanded and then the expectation over dX was taken. The procedure is reversed here. In light of the

2.3 THE DPE FOR A $\pi(t)$-DRIVEN PROCESS

transition rates (2.3.2), taking the expectation in (2.3.5) gives

$$\begin{aligned}J(x, t) = \max_{u}\{&e^{-rt}L(t, x, u)\, dt + o(dt) \\&+ J(x + b(t, x, u)\, dt, t + dt)(1 - (\lambda(x, u) + \mu(x, u))\, dt) \\&+ J(x + b(t, x, u)\, dt + a(t, x, u)\delta_1, t + dt)\lambda(x, u)\, dt \\&+ J(x + b(t, x, u)\, dt - a(t, x, u)\delta_2, t + dt)\mu(x, u)\, dt\}.\end{aligned} \quad (2.3.6)$$

The last three terms in (2.3.6) come from the expectation over dX. They correspond to no jump, a jump up by an amount δ_1, and a jump down by an amount δ_2, respectively. Taylor expanding all the terms in (2.3.6) and collecting them according to the order of dt gives

$$\begin{aligned}J(x, t) = \max_{u}\{&e^{-rt}L(t, x, u)\, dt + J(x, t) + J_t\, dt \\&- [\lambda(x, u) + \mu(x, u)]J(x, t)\, dt + J_x(x, t)b(t, x, u)\, dt \\&+ J(x + a(t, x, u)\delta_1, t)\lambda(x, u)\, dt \\&+ J(x - a(t, x, u)\delta_2, t)\mu(x, u)\, dt + o(dt)\}.\end{aligned} \quad (2.3.7)$$

Dividing by dt and letting $dt \to 0$ gives the DPE

$$\begin{aligned}0 = \max_{u}\{&e^{-rt}L(t, x, u) - [\lambda(x, u) + \mu(x, u)]J(x, t) \\&+ J_x(x, t)b(t, x, u) + J(x + a(t, x, u)\delta_1, t)\lambda(x, u) \\&+ J(x - a(t, x, u)\delta_2, t)\mu(x, u)\} + J_t.\end{aligned} \quad (2.3.8)$$

Observe that (2.3.8) is a partial differential–difference equation, since it involves partial derivatives of $J(x, t)$ as well as differences in the x argument. The term involving $J_x(x, t)$ in (2.3.8) comes from the deterministic part of the dynamical equation (2.3.1). That is, in the absence of noise, (2.3.1) becomes $dX = b(t, X, u)\, dt$; the portion of (2.3.8) involving $J_x(x, t)$ can be viewed as a "drift" term.

Exercise 2.3.1

Consider the special case in which

$$\begin{aligned}b(t, x, u) &\equiv 0, \qquad a(t, X, u) \equiv 1, \\\delta_1 &= 1, \qquad \lambda(x, u) = \lambda_0, \text{ a constant,} \\\mu(x, u) &\equiv 0.\end{aligned} \quad (2.3.9)$$

What can be said about the process $X(t)$? What happens to the DPE?

If (2.3.8) is Taylor expanded, with only one term kept, it becomes

$$0 = J_t + \max_u \{e^{-rt}L(t, x, u) + J_x(x, t)[b(t, x, u) \\ + a(t, x, u)\delta_1\lambda(x, u) - a(t, x, u)\delta_2\mu(x, u)]\}. \quad (2.3.10)$$

Equation (2.3.10) is the "deterministic limit" of (2.3.8), since it is the DPE corresponding to the deterministic dynamics

$$dX/dt = b(t, X, u) + a(t, X, u)[\delta_1\lambda(X, u) - \delta_2\mu(X, u)] \quad (2.3.11)$$

and objective functional

$$J(x, t) = \max_u \left[\int_t^T e^{-rs}L(s, X(s), u(s))\, ds + \varphi(X(T))\,|\,X(t) = x \right]. \quad (2.3.12)$$

Exercise 2.3.2

Let $\delta_1, \delta_2 \to 0$, $\lambda, \mu \to \infty$ so that

$$\delta_1\lambda(x, u) \to \eta_1(x, u), \text{ finite}; \\ \delta_2\mu(x, u) \to \eta_2(x, u), \text{ finite}. \quad (2.3.13)$$

To what *deterministic* control problem does this correspond? How should one interpret the limits in (2.3.13)? [Hint: Think of δ_1 as a quantity of resource discovered at rate $\lambda(x, u)$.]

Keeping two terms in the expansion of (2.3.8) gives a second-order partial differential equation as an approximation for (2.3.8). This equation is

$$0 = J_t + \max_u \{e^{-rt}L(t, x, u) \\ + J_x(x, t)[b(t, x, u) + a(t, x, u)\{\delta_1\lambda(x, u) - \delta_2\mu(x, u)\}] \quad (2.3.14) \\ + \tfrac{1}{2}J_{xx}[(a(t, x, u)\delta_1)^2\lambda(x, u) + (a(t, x, u)\delta_2)^2\mu(x, u)]\}.$$

Equation (2.3.14) is the "diffusion limit" of (2.3.8).

Exercises 2.3.3

To what $W(t)$-driven problem does (2.3.11) correspond? How would one interpret the diffusion coefficient? Is this a good approximation to (2.3.8)?

Bibliographic Notes

The books by Karlin and Taylor (1977, 1981) and Feller (1968, 1971) provide good introductions to the jump processes and are essential background material. DPEs such as (2.3.8) are extremely difficult to solve and there is little or no general theory for such equations, in

2.4. THE DPE WITH PARAMETER UNCERTAINTY

contrast to the partial differential equations derived in the previous section. The use of "diffusion approximations" such as (2.3.14) to deal with equations such as (2.3.8) is common in physics and chemistry [see, for example, the paper by Moyal (1949) or book by Van Kampen (1981)]. The validity of such equations is another issue and has recently been investigated by Knessl et al. (1984) and Matkowsky et al. (1984).

2.4 THE DYNAMIC PROGRAMMING EQUATION WITH PARAMETER UNCERTAINTY

The results in Sections 2.2 and 2.3 are useful, but they do not encompass all cases. For example, what happens if $a(t, x, u)$, $b(t, x, u)$, $\lambda(x, u)$, or $\mu(x, u)$ are known imprecisely? A typical case would be one in which any of these coefficients contains an unknown parameter and the parameter has a probability distribution attached to it.

Such problems are different from the ones in the previous section because information about the parameter becomes important when formulating the DPE. These problems are called "problems with imperfect information." It is always possible to enlarge the state space in such problems, but this approach rarely leads to a solution.

The results obtained in this section are used in Section 4.3 on the exploitation of an uncertain resource with learning, 5.3 on surveys of fish stocks, and 6.3 on managing a fluctuating population in the case of parameter uncertainty.

The general theory about the DPE with parameter uncertainty is quite sketchy. It is treated, to some extent, in Section 2.8. In this section, the basic ideas of the DPE with parameter uncertainty are illustrated through two examples.

Example 2.4.1: The Two Fishing Ground Problem

Consider two fishing grounds, in which a parameter λ_i is proportional to the density of fish in ground i. Assume that λ_i is not known exactly, but has a gamma distribution with parameter v_i and α_i. How should k fishing vessels be allocated in time to optimize the expected harvest? This problem is fundamentally different from those in the preceding two sections because *information* is gained as fish are caught. In particular, the rate at which fish are caught provides information about the unknown parameter λ_i. The state variable here, $X_i(t)$, is the total catch on ground i up to time t. The simplest model for the state dynamics is the Poisson process, so that

$$dX_i = d\pi_i.$$

Here $d\pi_i = 1$ with probability $\lambda_i \, dt$ and $d\pi_i = 0$ with probability $1 - \lambda_i \, dt$. If k vessels are fishing independently, then in the period $(0, t)$,

$$\Pr\{k \text{ vessels catch } n \text{ schools of fish in } (0, t) | \lambda\} = e^{-k\lambda t}(k\lambda t)^n / n!. \quad (2.4.1)$$

There are a number of assumptions underlying (2.4.1). Some of the most crucial ones are that (1) the search for fish is a random process, (2) the fishermen search independently, and (3) depletion of the stock can be ignored. A discussion of some of these assumptions is found in Mangel and Clark (1983). Since λ is not known, the true distribution of the catch is found by integrating (2.4.1) against the distribution on λ. Once a school of fish is caught, the method of Bayesian updating can be used to modify the distribution on λ. According to Bayes' theorem,

$$\Pr\{\lambda \in (\bar{\lambda}, \bar{\lambda} + d\bar{\lambda}) | k \text{ vessels caught } n \text{ schools of fish in } (0, t)\}$$

$$= \frac{\exp(-k\bar{\lambda}t)[(k\bar{\lambda}t)^n / n!] \exp(-\alpha\bar{\lambda})\bar{\lambda}^{v-1}\alpha^v / \Gamma(v)}{\int_0^\infty [\exp(-k\bar{\lambda}t)/n!][(k\bar{\lambda}t)^n / \Gamma(v)] \exp(-\alpha\bar{\lambda})\bar{\lambda}^{v-1}\alpha^v \, d\bar{\lambda}} d\bar{\lambda} \quad (2.4.2)$$

$$= \frac{\exp[-(\alpha + kt)\bar{\lambda}](\alpha + kt)^{v+n}\bar{\lambda}^{v+n-1}}{\Gamma(v + n)} d\bar{\lambda}. \quad (2.4.3)$$

Thus the posterior density for λ is again a gamma density, with updated parameters $\alpha + kt$ and $v + n$. The coefficient of variation of this density is $1/\sqrt{v + n}$, so that as long as at least one school of fish is caught information is gained (since $1/\sqrt{v_i + n} < 1/\sqrt{v_i}$).

In the statistical literature, the Poisson equation (2.4.1) is called a *prior density*, and the gamma density is called a *conjugate prior*. Statisticians have tabulated pairs that integrate out nicely; see, for example, the book by DeGroot (1970). A conjugate prior such as the gamma density just used is called an *informative prior*, since assumptions about the shape and location of the distribution on λ are made. They are also *noninformative priors* that require no such assumptions. They are discussed in the literature on statistical decision theory. For simplicity of exposition, they are not included with this problem.

To start the fishing optimization problem, one allocates boats over the two fishing grounds for one period of fishing of length T. At the end of this first period, vessels are reallocated among the two fishing grounds, depending on the updated probabilities. The process is repeated during the second period, and a further reallocation is undertaken if necessary. Suppose there are M such periods of length T constituting the total annual fishing season. As the simplest objective function, take the total seasonal catch. By making this choice, one ignores all costs incurred in fishing and in moving vessels from one fishing ground to another. Let $k_j^{(i)}$ be the number of vessels on

2.4. THE DPE WITH PARAMETER UNCERTAINTY

fishing ground j in period i. The optimization problem is to determine the allocation of vessels $k_1^{(i)}$, $k_2^{(i)}$ for $i = 1, 2, \ldots, M$ that maximizes the expected catch. What is the DPE for this problem? To find it, consider the situation in which the updated gamma distributions v_i, α_i have just been obtained and the reallocation decision is about to be made.

Now, if there are k_i vessels on fishing ground i, the expected catch given λ_i is $k_i \lambda_i T$, where T is the length of a fishing period. Averaging over λ_i gives $k_i v_i T/\alpha_i$. Let $J_n(v, \alpha)$ denote the maximum expected gain in the rest of the season when n periods of fishing remain and the current values of the parameters are $v = (v_1, v_2)$ and $\alpha = (\alpha_1, \alpha_2)$. When only one period of fishing remains, $J_1(v, \alpha)$ is found simply:

$$J_1(v, \alpha) = \max_{k_1}\{k_1 T v_1/\alpha_1 + (K - k_1) T v_2/\alpha_2\}, \tag{2.4.4}$$

where K is the total number of vessels available. The value of k_1 that solves (2.4.4) is

$$k_1^* = \begin{cases} K, & \text{if } v_1/\alpha_1 \geq v_2/\alpha_2, \\ 0, & \text{if } v_1/\alpha_1 < v_2/\alpha_2. \end{cases} \tag{2.4.5}$$

This solution indicates that if fishing ground 1 is better than fishing ground 2 (on the average), all effort should be put into fishing ground 1. If fishing ground 2 is better, all effort should be put into it. This simple solution arises for two reasons. First, the objective function $J_1(v, \alpha)$ is linear in k_1. Second, any information gained in this last period of fishing is valueless, since further fishing does not occur. This (trivial) solution is obtained because of the linearity of (2.4.4).

When more than one period remains, this simple rule breaks down because searching and harvesting provide information about the density of fish. Consider $J_2(v, \alpha)$ explicitly. Rewrite it as [with $k_1(2)$ the number of vessels on fishing ground 1 with 2 periods remaining]

$$J_2(v, \alpha) = \max_{k_1(2)}\{k_1(2) T v_1/\alpha_1 + [K - k_1(2)] T v_2/\alpha_2$$

$$+ \sum_{m_1=0}^{\infty} \sum_{m_2=0}^{\infty} \Pr\{k_1(2) \text{ vessels find } m_1 \text{ schools of fish}\}$$

$$\times \Pr\{K - k_1(2) \text{ vessels find } m_2 \text{ schools of fish}\} J_1(v + m, \alpha + k(2)T)\}. \tag{2.4.6}$$

Here $m = (m_1, m_2)$ and $k(2) = (k_1(2), K - k_1(2))$.

The first two terms in (2.4.6) correspond to the expected catch in the second to the last period of fishing. The double summation in (2.4.6) is the expectation over the last period, given the discoveries of the next to the last period.

The probability in (2.4.6) can be calculated explicitly by averaging over (2.4.1). If $k_1(2) \neq 0$,

$$\Pr\{k_1(2) \text{ vessels find } m_1 \text{ schools of fish}\}$$

$$= \int_0^\infty \Pr\{m_1 | \lambda, k, (2)\} \gamma(\lambda; v_1, \alpha_1) \, d\lambda \qquad (2.4.7)$$

$$= \frac{k_1(2)^{m_1}}{m_1!} \frac{\alpha_1^{v_1}}{(\alpha_1 + k_1(2)T)^{m_1 + v_1}} \frac{\Gamma(m_1 + v_1)}{\Gamma(v_1)}. \qquad (2.4.8)$$

Of course, if $k_1(2) = 0$, then $m_1 = 0$ with probability 1. In the literature of decision theory, (2.4.6) is called a *preposterior expectation.* It involves data from the next to the last period, as well as an optimal decision over the last period. Note that one cannot determine the action in the last period in advance; the data from the next to the last period are needed to make that decision. Equation (2.4.6) is easily solved on a microcomputer.

What about $J_3(v, \alpha)$? To find it, one must compute $J_2(v, \alpha)$ and $J_1(v, \alpha)$ for *all* possible values of v and α greater than the initial values. This is a difficult task; it illustrates the so-called "curse of dimensionality" of dynamic programming.

Exercise 2.4.2

Find the dynamic programming equation that $J_3(v, \alpha)$ must satisfy. Are there any special assumptions that could be used to help find the solution?

Another complication arises if one tries to include depletion in the model of fishing. In this case the distribution on λ as well as the parameters change, and the DPE is even more complicated. One approach to solving this problem is found in Mangel and Clark (1983), in which other aspects are also considered.

Example 2.4.3: The Two-Armed Bandit Problem

This example has a long history associated simultaneously with medical trials and slot machines. Consider two gambles G_1 and G_2 which pay off according to the following rules:

(i) G_1 pays α_1 with probability p_1, nothing with probability $1 - p_1$;
(ii) G_2 pays α_2 with probability p_2, nothing with probability $1 - p_2$.

In the simplest case, it is assumed that p_2 is known with certainty but that p_1 is not known exactly (its probability density will be specified below).

Consider the situation in which N gambles are available. Assume that the cost of G_1 and G_2 are the same. The problem is to determine the optimal allocation of the total number of gambles N between G_1 and G_2.

2.4. THE DPE WITH PARAMETER UNCERTAINTY

If no information on the parameters is obtained as gambles occur, then the answer would be to put all N gambles where $\alpha_i p_i$ (the expected gain from a single gamble) is highest. However, each gamble G_1, whether successful or not, provides information about p_1.

The natural question is, how does one model learning and incorporate information into this problem? First, note that if p_1 were given, then the probability of k wins in m gambles is given by

$$\Pr\{k \text{ wins in } m \text{ gambles } G_1 | p_1\} = \binom{m}{k} p_1^k (1 - p_1)^{m-k}. \quad (2.4.9)$$

The next step is to find a "good" conjugate prior for p_1. Suppose that $f(p_1)$ is the prior density for p_1. Applying Bayes theorem to a situation in which k of m gambles were successful gives

$$\Pr\{p_1 \in (p, p + dp) | k \text{ wins in } m \text{ gambles } G_1\}$$

$$= \frac{f(p)\binom{m}{k} p^k (1-p)^{m-k}}{\int_0^1 f(p)\binom{m}{k} p^k (1-p)^{m-k} dp} dp. \quad (2.4.10)$$

Inspection of (2.4.10) shows that a good choice for $f(p_1)$ is the beta density given by

$$f(p_1) = [\Gamma(\alpha + \beta)/\Gamma(\alpha)\Gamma(\beta)] p_1^{\alpha - 1} (1 - p_1)^{\beta - 1}. \quad (2.4.11)$$

With this choice, the denominator in (2.4.10) integrates exactly to give

$$\int_0^1 f(p) p^k (1-p)^{m-k} dp = \frac{\Gamma(\alpha + \beta)}{\Gamma(\alpha)\Gamma(\beta)} \frac{\Gamma(\alpha + k)\Gamma(\beta + m - k)}{\Gamma(\alpha + \beta + m)}. \quad (2.4.12)$$

Exercise 2.4.4

Show that if $f(p_1)$ is given by (2.4.11), then the mean and variance of p_1 are

$$E\{p_1\} = \alpha/(\alpha + \beta), \quad \text{Var}\{p_1\} = \frac{\alpha\beta}{(\alpha + \beta)^2(\alpha + \beta + 1)}. \quad (2.4.13)$$

The posterior density for p_1 given the information of k wins in m gambles is now

$$\Pr\{p_1 \in (p, p + dp) | k \text{ wins in } m \text{ gambles}\}$$

$$= \frac{\Gamma(\alpha + \beta + m)}{\Gamma(\alpha + k)\Gamma(\beta + m - k)} p_1^{k + \alpha - 1} (1 - p_1)^{\beta + m - k - 1}, \quad (2.4.14)$$

which is another beta density. The choice of (2.4.11) as the prior density allows one to keep the density on p_1 closed. That is, the prior is a beta density and so is the posterior—only the parameters have shifted. By assuming the density in (2.4.11), one introduces a special structure to the problem. This special structure allows one to obtain a DPE that is relatively easy to solve. To find the DPE, let $J_n(\alpha, \beta)$ be the maximum gain from the nth gamble onwards, given that the current values for the parameters in $f(p_1)$ are α and β. Then

$$J_N(\alpha, \beta) = \max\{\alpha_2 p_2, \alpha_1[\alpha/(\alpha + \beta)]\} \qquad (2.4.15)$$

is the expected gain from the gamble on the last gamble. To derive equation (2.4.15), observe that any information obtained on the last gamble is valueless, so that one simply chooses the gamble with the larger expected value. For G_2 the expected gamble is simply $\alpha_2 p_2$. For G_1 the expected value, conditioned on p_1, is $\alpha_1 p_1$; this expectation must then be averaged over the distribution on p_1. When this is done, one obtains (2.4.15).

Exercise 2.4.5

Derive the DPE for the remaining gambles.

Bibliographic Notes

The books by DeGroot (1970), Berger (1980), and Martz and Waller (1982) provide good introductions to Bayesian analysis as well as conjugate priors and their uses. The method of deriving DPEs by starting with the last period first is called "backward induction." Bertsekas (1976) and Aoki (1967) provide some examples of DPEs with *parameter uncertainty*, which is also known as imperfect information. Example 2.4.1 is based on the work of Mangel and Clark (1983). Many of these Bayesian techniques for analyzing information have been used to study foraging behavior in animals. The papers by Krebs, Kacelnik, and Taylor (1978) and Green (1980) and the books edited by Kamil and Sargent (1981), Krebs and Davies (1978), and Anderson, Turner, and Taylor (1979) provide good introductions to the foraging literature and the use of Bayesian methods in foraging theory. The two-armed bandit problem is an example of sequential allocation in experiments and has a large literature associated with it. Bather (1981) provides a good review of this literature.

2.5 FINDING OPTIMAL STOCHASTIC CONTROLS DIRECTLY

Recall that for the $W(t)$-driven process, the DPE is the partial differential equation

$$0 = J_t + \max_{u \in V}\{e^{-rt}L(s, x, u) + b(t, x, u)J_x + \tfrac{1}{2}a(t, x, u)J_{xx}\}. \qquad (2.5.1)$$

2.5. FINDING OPTIMAL STOCHASTIC CONTROLS DIRECTLY

It should be clear by now that solving such an equation is a formidable task. Any heuristic that can be used to get at a solution should be investigated.

This section contains a description of the method of Arkin et al. (1966). It is based on the following observation. In most control problems, one wishes to find the control as a function of the state, that is, $u^* = u^*(x)$. Under certain assumptions, (2.5.1) can be used to derive an equation for the optimal control $u^*(x)$ directly, so that by solving this equation the optimal control is found without finding $J(x, t)$ explicitly.

The results of this section are used in Section 4.5 on the price dynamics of and markets for exhaustible resources, 6.4 on the price dynamics of renewable resource markets, and 6.5 on the trade-off between stock and control fluctuations.

To illustrate the method in its simplest form, assume that

(1) $b(\cdot)$ and $L(\cdot)$ are functions of x and u only,
(2) $a(\cdot)$ is a function of x only,
(3) the discount rate $r = 0$,
(4) the boundary condition requires that $J(x, t)$ vanishes at two points l_1 and l_2.

The only crucial assumption is that $a(\cdot)$, $b(\cdot)$, and $L(\cdot)$ are independent of time. The rest of the assumptions can be relaxed if necessary.

With these assumptions, one can set $J_t \equiv 0$. The DPE (2.5.1) becomes

$$0 = \tfrac{1}{2}a(x)J_{xx} + \max_{u \in V}\{L(x, u) + b(x, u)J_x\}. \qquad (2.5.2)$$

Equation (2.5.2) is, in general, a nonlinear, second-order ordinary differential equation. *Assuming* that the optimal control is an internal control, we differentiate the term in braces and set it equal to zero. This gives

$$L_u(x, u) + b_u(x, u)J_x = 0. \qquad (2.5.3)$$

Equation (2.5.3) is one equation for the two unknowns $u(x)$ and J_x. Solving (2.5.3) for J_x gives $J_x = -L_u(x, u)/b_u(x, u)$. Differentiating (2.5.3) with respect to x and recalling that u is a function of x gives

$$L_{ux} + L_{uu}\frac{du}{dx} + J_{xx}b_u + J_x\left(b_{ux} + b_{uu}\frac{du}{dx}\right) = 0. \qquad (2.5.4)$$

Equations (2.5.3) and (2.5.4) are two equations for the three unknowns $u(x)$, J_x, and J_{xx}. Solving (2.5.4) for J_{xx} gives

$$J_{xx} = \left[-J_x\left(b_{ux} + b_{uu}\frac{du}{dx}\right) - L_{ux} - L_{uu}\frac{du}{dx}\right]\bigg/ b_u. \qquad (2.5.5)$$

The DPE is the third equation needed to complete the set, that is, to give three equations for the three unknowns.

Substituting (2.5.3) and (2.5.5) in the DPE (2.5.2) [understanding below that $u = u^*(x)$, the optimal feedback control] gives an equation for $u(x)$:

$$0 = \tfrac{1}{2}a(x)\left[\frac{L_u}{b_u}\left(b_{ux} + b_{uu}\frac{du}{dx}\right) - L_{ux} - L_{uu}\frac{du}{dx}\right]\bigg/ b_u$$

$$+ L(x, u) + b(x, u)(-L_u/b_u). \qquad (2.5.6)$$

Equation (2.5.6) can be rearranged to obtain an ordinary differential equation for $u(x)$:

$$\frac{du}{dx}\left(\frac{L_u}{b_u}b_{uu} - L_{uu}\right) = \frac{2}{a(x)}\left[\frac{bL_u}{b_u} - L\right] + L_{ux} - \frac{L_u b_{ux}}{b_u}. \qquad (2.5.7)$$

Assuming that the coefficient of du/dx is nonzero, (2.5.7) can be rewritten as

$$du/dx = H(x, u). \qquad (2.5.8)$$

The solution of (2.5.8) is the feedback control, obtained without ever using J explicitly.

Exercise 2.5.1

Derive an equation for $u(x)$ if the diffusion term $a(x, u)$ also depends on the control.

To solve (2.5.7), one must specify an initial condition $u(x) = u_0$ at $x = x_0$. Arkin et al. (1966) does not provide a method for this. However, assumption (4) does provide a method. Assume that $u(x_0) = u_0$; then the solution of (2.5.8) is the feedback rule $u = u(x; u_0, x_0)$. In light of (2.5.3), J_x is given by

$$J_x = -L_u(x, u(x; u_0, x_0))/b_u(x, u(x; u_0, x_0)). \qquad (2.5.9)$$

Integrating (2.5.9) gives

$$J(x) = \int_x^{l_2} [L_u(s, u(s; u_0, x_0))/b_u(s, u(s; u_0, x_0))]\, ds. \qquad (2.5.10)$$

Hence $J(l_2) \equiv 0$, and applying the other boundary condition gives an implicit equation for u_0:

$$0 = \int_{l_1}^{l_2} \frac{L_u(s, u(s; u_0, x_0))}{b_u(s, u(s; u_0, x_0))}\, ds \qquad (2.5.11)$$

Exercise 2.5.2

Consider a specific problem for which

$$L(x, u) = 1 - cu^m, \qquad (2.5.12)$$

$$b(x, u) = u, \quad \alpha(x - u) \quad \text{or} \quad \alpha x u/(u + \beta), \qquad (2.5.13)$$

and $a(x, u)$ is a constant. How far can one go analytically in finding the feedback control $u^*(x)$? Be sure to include any strange behavior of the solution in your discussion.

Bibliographic Notes

The paper of Arkin *et al.* (1966) seems to stand alone in the description and use of this technique. The method once again requires the solution of nonlinear ordinary differential equations.

2.6 CONTROL WITH PROBABILITY CRITERIA

One kind of problem that is often important for resource management is understanding the balance between costs and chances of undesirable events. One way of characterizing an undesirable event is by the *probability* of its occurrence. In such a case, the dynamic programming formalism leads to *probabilistic programming*, as it is often called [e.g., Vajda (1970)]. These are dynamic programming problems with probability criteria.

The results described in this section are used in Section 6.2 on harvesting a randomly fluctuating population and 6.5 on the trade-off between stock and control fluctuations.

The idea is best illustrated with a simple example from which more complicated ones can be patterned. So, to begin, start with the SDE for a single variable $X(t)$:

$$dX = b(X, u)\, dt + \sqrt{a(X, u)}\, dW, \qquad (2.6.1)$$

$$X(0) = x_0, \quad l_1 < x_0 < l_2. \qquad (2.6.2)$$

Here $b(X, u)$ and $a(X, u)$ are presumed to be known and u is the control variable.

For the objective function, first consider the probability

$$p(t, T, x) = \min_{u} \Pr\{X(\tau) \text{ exited } (l_1, l_2) \text{ by } T \,|\, X(t) = x\}. \qquad (2.6.3)$$

Exercise 2.6.1

Describe a situation in which (2.6.3) would be a reasonable objective function.

To derive a DPE for $p(t, T, x)$, observe that in a time interval dt the process jumps to the point $x + dX$, where dX is normally distributed. The mean of the jump is $b(x, u)\, dt + o(dt)$, and the variance of the jump is $a(x, u)\, dt + o(dt)$. Clearly $p(t, T, x)$ is then the same as the average of $p(t + dt, T, x + dX)$ over dX. This observation gives

$$p(t, T, x) = \min_u \mathrm{E}_{dX}\{p(t + dt, T, x + dX)\}. \tag{2.6.4}$$

Equation (2.6.3) leads directly to the dynamic programming equation

$$\frac{\partial p}{\partial t} + \min_u \left\{ b(x, u) \frac{\partial p}{\partial x} + \frac{1}{2} a(x, u) \frac{\partial^2 p}{\partial x^2} \right\} = 0. \tag{2.6.5}$$

What about boundary and initial or final conditions? These are $p(t, T, l_i) = 1$ for $i = 1, 2$ and $p(T, T, x) = 0$ if $l_1 < x < l_2$. In general, (2.6.5) must be solved by numerical methods. Although it is a hard equation to solve, (2.6.5) provides an explicit way of characterizing the probability of exit from the interval (l_1, l_2).

A second way to characterize the exit from an interval is by the *mean exit time*. Let $\bar{T}(x, u)$ be the mean exit time from (l_1, l_2), given that $X(0) = x$ and a control law $u(x)$, and let

$$\bar{T}(x) = \max_u \bar{T}(x, u). \tag{2.6.6}$$

That is, minimizing the probability of exit and maximizing the time until exit are, to some extent, analogous problems. One can derive an equation for $\bar{T}(x)$ by repeating the argument that lead to (2.6.4). When this is done, $\bar{T}(x)$ is shown to satisfy the equation

$$-1 = \max_u \left\{ b(x, u) \frac{\partial \bar{T}}{\partial x} + \frac{a(x, u)}{2} \frac{\partial^2 \bar{T}}{\partial x^2} \right\} \tag{2.6.7}$$

with the boundary conditions $\bar{T}(l_i) = 0$.

Exercise 2.6.2

Derive (2.6.7).

Example 2.6.3

Consider a simple example in which $b(x, u) = x + u$, $a(x, u) = \varepsilon$. Equation (2.6.7) becomes

$$-1 = \frac{\varepsilon}{2} \frac{\partial^2 \bar{T}}{\partial x^2} + x \frac{\partial \bar{T}}{\partial x} + \max_u \left\{ u \frac{\partial \bar{T}}{\partial x} \right\}. \tag{2.6.8}$$

2.6. CONTROL WITH PROBABILITY CRITERIA

For this problem to make sense, the control u must be bounded, say, $-u_m \leq u \leq u_m$. The optimal control is then given by

$$u^* = u_m \, \text{sgn}(\partial \bar{T}/\partial x) = u_m(\partial \bar{T}/\partial x)/|\partial \bar{T}/\partial x|. \tag{2.6.9}$$

Substituting (2.6.9) into (2.6.8) gives the following nonlinear equation for $\bar{T}(x)$:

$$-1 = \frac{\varepsilon}{2}\frac{\partial^2 \bar{T}}{\partial x^2} + x\frac{\partial \bar{T}}{\partial x} + u_m\left(\frac{\partial \bar{T}}{\partial x}\right)^2 \bigg/ \left|\frac{\partial \bar{T}}{\partial x}\right|. \tag{2.6.10}$$

Exercise 2.6.4

Suppose that $l_1 = -1$, $l_2 = 1$; write a computer code to calculate $\bar{T}(x)$ if $u \equiv 0$ (the uncontrolled case). Then write a code to compute the solution of (2.6.10) and compare the two results.

Example 2.6.5: Enhancement of an Idealized Fish Population

The following example is a caricature of enhancement programs for biological populations. It is a nice example that can be solved explicitly. Assume that the population of interest has a natural, stable steady state at P_0 and that this state is enhanced to a level $P_0 + \alpha E$ by expenditure of effort E at cost $c(E)$ (either by stocking with young or by removing predators). Let $X(t) = P(t) - (P_0 + \alpha E)$ denote the deviation from the steady state at level E. Assume that a linear approximation around the steady state is valid so that $X(t)$ satisfies the linear SDE

$$dX = -\beta X + \sqrt{a}\, dW, \tag{2.6.11}$$

where β and a are constants.

The steady-state density for X can be shown to be

$$q_s(x) = \sqrt{\beta/\pi a}\, \exp(-\beta x^2/a). \tag{2.6.12}$$

Recall that (2.6.11) is the SDE for the Ornstein–Uhlenbeck process and that the time-dependent probability density for the process can also be found. Since this example is meant to be illustrative, the simpler equation (2.6.12) is used.

Exercise 2.6.6

Derive (2.6.12) by using a forward equation (Section 1.1.2) and setting its time derivative equal to zero.

A typical managerial question might concern the chance that the population ever falls below a given value, say, $P_{\min}$, for a given level of enhancement. To answer such a question, one can proceed as follows.

We define $q(p)$ by

$$q(p)\,dp = \Pr\{\text{steady-state population level is in } (p, p+dp)\}$$
$$= \Pr\{\text{steady-state value of deviation is in} \qquad (2.6.13)$$
$$(p - P_0 - \alpha E, p - P_0 - \alpha E + dp)\}.$$

Using (2.6.12) to find $q(p)$ gives

$$q(p) = \sqrt{\beta/\pi a}\, \exp\{(-\beta/a)[p - (P_0 + \alpha E)]^2\}. \qquad (2.6.14)$$

Equation (2.6.14) provides an explicit way of computing the probability density for the deviation from the enhanced level $P_0 + \alpha E$.

Suppose that the goal is to determine a constant (stationary) value of E to maximize the probability that the population is above the minimal level $P_{\min}$ at minimal relative cost. This suggests considering the functional

$$J = \int_{P_{\min}}^{\infty} q(p)\,dp - c(E). \qquad (2.6.15)$$

Setting $y = p - (P_0 + \alpha E)$, (2.6.15) can be rewritten as

$$J(E) = \int_{P_{\min} - (P_0 + \alpha E)}^{\infty} \sqrt{\beta/\pi a}\, \exp(-\beta y^2/a)\,dy - c(E). \qquad (2.6.16)$$

To find the optimal level of effort, we set the derivative $J'(E)$ equal to zero. This gives the following nonlinear equation for E:

$$0 = J'(E) = \sqrt{\beta/\pi a}\, \exp\{(-\beta/a)[P_{\min} - (P_0 + \alpha E)]^2\}\alpha - c'(E). \qquad (2.6.17)$$

Exercise 2.6.7

Assume that $c(E) = E^n$ or $c(E) = \exp\{wE^n\}$ for fixed n, w. Determine the optimal level of effort and the steady-state probability of the population being above $P_{\min}$.

Bibliographic Notes

Control with probability criteria is discussed by Wonham (1969), who derives the appropriate DPE as well as discusses some methods of solution. Numerical methods for such problems are discussed by van Mellaert and Dorato (1972). Mendelssohn (1979) considers some ecological problems in which control with probability criteria is important. Example 2.6.5 is motivated by the work of Foerster and Ricker (1941). This work is discussed by Larkin (1977), who points out that the work was not really followed upon and is still a potentially promising research area.

2.7 MYOPIC BAYES STOCHASTIC PROBLEMS AND NONLINEAR PROGRAMMING

A main problem with stochastic DPEs is the "curse of dimensionality," that is, for the calculation of J_{n-1} (in the discrete case) or $J(t - dt)$ (in the continuous case), the entire history $J_n, J_{n+1}, J_{n+2}, \ldots, J_N$ or $J(s)$, $t \leq s \leq T$, must be computed and stored. For most resource problems, which involve large numbers of periods, this creates real problems with storage and memory in the computer used to solve the problem.

One way around this dilemma is to use a "Myopic Bayes" strategy. According to this strategy, an N-period discrete problem is converted into a two-stage problem. The first stage contains j periods, and the second stage contains $N - j$ periods. The functional now requires only two calculations. For a continuous time problem, the first stage has length τ, the second stage length $T - \tau$. These ideas are best illustrated by a sequence of (contrived) examples. More realistic examples will come in later sections. The ideas presented here are used in Section 4.3 on the utilization of an exhaustible resource with learning, 5.3 on surveys of fish stocks, and 6.2 on harvesting a randomly fluctuating population.

Example 2.7.1

Reconsider Example 2.4.1, which was the problem concerning two fishing grounds. Equation (2.4.6) is the DPE of interest. With suitable modification, it can be considered a myopic Bayes DPE. To do this, divide the fishing season of M weeks into two periods of $M/2$ weeks each. Then T in (2.4.6) is replaced by $T(M/2)$ and $J_1(v + m, \alpha + k(2)T)$ is interpreted as the optimal gain in the last period. With these modifications (2.4.6) is a myopic Bayes DPE.

The term "myopic Bayes" is quite an accurate description. The procedure is Bayesian in that information is obtained and used in the Bayesian way. The procedure is myopic because it assumes that information is obtained only once. It is also clear that (2.4.6), when written as a two-stage problem, is simply a problem in nonlinear programming.

Exercise 2.7.2

Reconsider Example 2.4.1 for the case of a single fisherman who can visit only one of the two grounds in each week. First derive the full DPE for this problem. Second, derive the myopic Bayes version of the DPE. Third, code the myopic Bayes version of the DPE and examine the numerical results.

Example 2.7.3

Next, consider a model for a resource that is continuously distributed (e.g., a model for oil, water, or mineral deposits). In such cases the discovery of the resource is often costly, but the known deposit has a value of its own (the value of "known reserves"). In particular, assume that in a period of length t the discovery process can be characterized by

$$\begin{aligned}&\Pr\{\text{amount of resource found in } (0, t) \text{ is in the}\\&\quad\text{interval } (x, x + dx) \text{ with } k \text{ searchers} | \lambda\}\\&= [x\lambda/(kt)^2] \exp[(-\lambda/2)(x/kt)^2] \, dx.\end{aligned} \qquad (2.7.2)$$

The density in (2.7.2) is sometimes called the Rayleigh density. In (2.7.2), λ is a measure of the ease of finding the resource. The distribution function is $1 - \exp[-\lambda x^2/2(kt)^2]$.

Exercise 2.7.4

If X has density (2.7.2), find $E\{X\}$.

Assume that λ is unknown and has a gamma density with parameters v and α. The unconditional density for the amount of resource discovered is then given by

$$\begin{aligned}&\Pr\{\text{amount found is between } (x, x + dx) | k, (0, t)\}\\&= [x/(kt)^2] \int_0^\infty \lambda \exp[(-\lambda/2)(x/kt)^2][\alpha^v \exp(-\alpha\lambda)/\Gamma(v)]\lambda^{v-1} \, d\lambda\\&= [x/(kt)^2][\alpha^v/\Gamma(v)]\{\Gamma(v + 1)/[\alpha + \tfrac{1}{2}(x/kt)^2]^{v+1}\}\\&= [vx/(kt)^2][\alpha^v/(\alpha + x^2/2(kt)^2)^{v+1}].\end{aligned} \qquad (2.7.3)$$

Consider now the problem of Bayesian updating on the value of λ. Given discovery information in $(0, t)$, the posterior density on λ is

$$\Pr\{\lambda \in (\bar\lambda, \bar\lambda + d\bar\lambda) | k \text{ searchers found } x \text{ in } (0, t)\}$$
$$= \frac{\exp[-\bar\lambda(\alpha + \tfrac{1}{2}(x/kt)^2)]\bar\lambda^v(\alpha + x^2/2(kt)^2)^{v+1}}{\Gamma(v + 1)} \, d\bar\lambda. \qquad (2.7.4)$$

Suppose that each searcher has a fixed cost of search c_1 and a variable cost of search c_2. For the two-period problem [with k_1 equal to the number of

2.7. MYOPIC BAYES STOCHASTIC PROBLEMS

searchers in $(0, t)$ and k_2 equal to the number of searchers in (t, T)], the DPE is

$$J = \max_{k_1}\left\{pk_1 t \, E\{1/\sqrt{\lambda}\} - c_1 k_1 - c_2 k_1 t \right.$$
$$+ \int_0^\infty \{[vx/(k_1 t)^2]\{\alpha^v/[\alpha + x^2/2(k_1 t)^2]^{v+1}\}$$
$$\left. \times \left\{\max_{k_2}[pk_2(T-t) \, E'\{1/\sqrt{\lambda}\} - c_1 k_2 - c_2 k(T-t)]\right\} dx \right\}. \quad (2.7.5)$$

In (2.7.5) the first expectation is found using the original gamma density, so that

$$E\left\{\frac{1}{\sqrt{\lambda}}\right\} = \int_0^\infty \lambda^{-1/2} \frac{\alpha^v \lambda^{v-1} e^{-\alpha\lambda}}{\Gamma(v)} d\lambda = \frac{\Gamma(v - 1/2)}{\Gamma(v)} \sqrt{\alpha}. \quad (2.7.6)$$

The second expectation in (2.7.5) is found using the posterior density (2.7.4), so that

$$E'(1/\sqrt{\lambda}) = \int_0^\infty \frac{\exp[-\lambda(\alpha + \tfrac{1}{2}(x/k_1 t)^2)]\lambda^{v+1/2-1}}{\Gamma(v+1)} \left(\alpha + \frac{x^2}{2(k_1 t)^2}\right)^{v+1} d\lambda$$
$$= [\Gamma(v + 1/2)/\Gamma(v + 1)][\alpha + x^2/2(k_1 t)^2]^{1/2}. \quad (2.7.7)$$

Substituting (2.7.6) and (2.7.7) in (2.7.5) gives the nonlinear programming problem

$$J = \max_{k_1}\left\{pk_1 t \sqrt{\alpha}[\Gamma(v - 1/2)/\Gamma(v)] - c_1 k_1 - c_2 k_1 t \right.$$
$$+ \int_0^\infty \{[vx/(k_1 t)^2][\alpha^v/(\alpha + 1/2(x/k_1 t)^2)^{v+1}]\}$$
$$\times \left\{\max_{k_2}\{pk_2(T-t)[\Gamma(v + 1/2)/\Gamma(v + 1)][\alpha + 1/2(x/k_1 t)^2]^{1/2}\right.$$
$$\left.\left. - c_1 k_2 - c_2 k_2(T-t)]\right\} dx \right\}. \quad (2.7.8)$$

The stochastic dynamic programming problem is now converted to a deterministic two-step nonlinear programming problem.

Bibliographic Notes

Bertsekas (1976, pp. 214–215) discusses the two-stage problem in more detail. Koblin (1971) calls the myopic approach the "class of two-stage stochastic programming problems," and discusses solutions as well as provides examples. Ludwig and Walters (1981, 1982) and Walters (1981) show how myopic Bayes procedures can be used in salmon management. They also compare the myopic procedures with optimal ones.

2.8 VALUES, STRATEGIES, AND POLICY ITERATION

This section contains a discussion of certain philosophical issues concerning the modeling of stochastic resource systems formulation of DPEs and some results on methods for obtaining solutions of DPEs.

The first point is the separation of strategies (i.e., decision rules) and values (the return from those decision rules). The point is best illustrated by an example, so we reconsider Example 2.4.1 (the problem of two fishing grounds) when there is only one fisherman and the fishing season consists of only two periods. As a problem in stochastic dynamic programming, this particular problem is not especially interesting. It does, however, allow one to discuss the philosophical and modeling issues as clearly as possible. The DPE derived in Section 2.4 simplifies as follows. Let $J_1(v_1, \alpha_1, v_2, \alpha_2)$ denote the maximum expected return in the last period of fishing, given that v_i and α_i are the current values of the parameters of the gamma density characterizing λ on ground i. Equation (2.4.4) is replaced by

$$J_1(v_1, \alpha_1, v_2, \alpha_2) = \max_i (v_i/\alpha_i)T. \tag{2.8.1}$$

Here T is the length of the fishing period, and it is assumed that the fisherman can go to only one of the two grounds in each period.

Let $J_2(v_1, \alpha_1, v_2, \alpha_2)$ denote the maximum expected gain when two periods of fishing remain and the current parameters are v_i and α_i. Instead of (2.4.6), one obtains the DPE analogous to (2.4.6):

$$\begin{aligned}&J_2(v_1, \alpha_1, v_2, \alpha_2) \\ &= \max_i \left\{ (v_i/\alpha_i)T + \sum_{n=0}^{\infty} p(n; v_i, \alpha_i) \right. \\ &\quad \left. \times J_1(v_1 + n\delta_{i1}, \alpha_1 + T\delta_{i1}, v_2 + n\delta_{i2}, \alpha_2 + T\delta_{i2}) \right\}. \end{aligned} \tag{2.8.2}$$

In this equation $p(n; v_i, \alpha_i)$ is the probability that the fisherman finds n schools on ground i when the values of the parameters are v_i and α_i; it is

2.8. VALUES, STRATEGIES, AND POLICY ITERATION

given by (2.4.8). In (2.8.2) δ_{ik} is the Kroenecker delta function, namely, $\delta_{ik} = 1$ if $i = k$ and zero otherwise.

Equations (2.8.1) and (2.8.2) are DPEs similar to those derived in the preceding sections. Buried in these equations are the true state of nature (the value of λ_i, $i = 1, 2$), the information (the number of schools of fish caught in the second to the last period of fishing), and the decision rule. For a problem as simple as this one, it is possible to combine these three facets (true state of nature, information, and decision rule) and still understand the nature of the solution. For more complicated problems it helps to explicitly separate the three components. The following approach can be used to separate the components of the problem in an understandable way.

In each period the fisherman has two strategies to choose from: to visit fishing ground 1 or to visit fishing ground 2. Let $\hat{V}_2(i)$ denote the *value* of visiting fishing ground i in the next to the last period of fishing. This value involves two terms. The first term is the expected catch in the next to the last period of fishing. Since fishing is a Poisson process with parameter λ_i, this expectation is $\lambda_i T$. The second term is the expected catch in the last period of fishing. To show how to compute it, consider $\hat{V}_2(1)$ explicitly, that is, the value to the fisherman of visiting fishing ground 1 first. In the next to last period of fishing, the fisherman catches n schools on fishing ground 1 with conditional probability $p(n, \lambda_1, T)$. In this model $p(n, \lambda_1, T)$ is the Poisson distribution with parameter $\lambda_1 T$. What is needed next is the explicitly stated decision rule. The decision rule can be explicitly derived as follows. If, in the next to last period of fishing, n schools of fish are found, then the posterior density on λ_1 is a gamma density with parameters $v_1 + n$ and $\alpha_1 + T$ and the posterior expected value of λ_1 is $(v_1 + n)/(\alpha_1 + T)$. Define n_1^* by

$$(v_1 + n_1^*)/(\alpha_1 + T) = v_2/\alpha_2. \qquad (2.8.3)$$

The interpretation of n_1^* is this: if fewer than n_1^* schools of fish are found in the first period, then fishing ground 2 has a better expected catch in the last period than fishing ground 1 and should be visited. If more than n_1^* schools of fish are found, then the reverse is true. For all practical purposes, the situation of n_1^* being an integer can be ignored.

Exercise 2.8.1

Practicality aside, what should be done if n_1^* is an integer?

The expected catch on either fishing ground in the last period is simply $\lambda_i T$. Consequently, conditioned on λ_1 and λ_2, the value of visiting fishing ground 1 first is

$$\hat{V}_2(1 | \lambda_1, \lambda_2) = \lambda_1 T + \sum_{n=0}^{n_1^*} p(n, \lambda_1, T) \lambda_2 T + \sum_{n=n_1^*+1}^{\infty} p(n, \lambda_1, T) \lambda_1 T. \qquad (2.8.4)$$

70 2. STOCHASTIC CONTROL AND DYNAMIC PROGRAMMING

To find $\hat{V}_2(1)$, one must average $\hat{V}_2(1|\lambda_1, \lambda_2)$ over λ_1 and λ_2. This gives

$$\hat{V}_2(1) = E_{\lambda_1 \lambda_2}\left\{\lambda_1 T + \sum_{n=0}^{n_1^*} p(n, \lambda_1, T)\lambda_2 T + \sum_{n=n_1^*+1}^{\infty} p(n, \lambda_1, T)\lambda_1 T\right\}. \quad (2.8.5)$$

Here $E_{\lambda_1 \lambda_2}$ denotes the expectation over both λ_1 and λ_2; it is found by integrating against each density.

The value of the strategy of visiting fishing ground 2 in the first period of fishing, $\hat{V}_2(2)$, is found in the same way. In particular, let n_2^* satisfy

$$v_1/\alpha_1 = (v_2 + n_2^*)/(\alpha_2 + T). \quad (2.8.6)$$

Then $\hat{V}_2(2)$ is given by

$$\hat{V}_2(2) = E_{\lambda_1 \lambda_2}\left\{\lambda_2 T + \sum_{n=0}^{n_2^*} p(n, \lambda_2, T)\lambda_1 T + \sum_{n=n_2^*+1}^{\infty} p(n, \lambda_2, T)\lambda_2 T\right\}. \quad (2.8.7)$$

The value functions $\hat{V}_2(1)$ and $\hat{V}_2(2)$ can now be used to determine which fishing ground should be visited first. That is, the ground with the higher value function should be visited first. (As before, the choice of ground in the second period depends upon the data obtained in the first period.) It appears that (2.8.5) and (2.8.7) are very different from the original DPE formulation (2.8.2).

To compare (2.8.5) and (2.8.7) with the DPE (2.8.2), it is useful to write (2.8.2) in a form similar to (2.8.5) or (2.8.7). To do this, let $V_2(i)$ be the maximum expected gain from the last two periods of fishing, subject to the strategy of visiting fishing ground i in the next to last period. From (2.8.2), it follows that

$$V_2(1) = E_{\lambda_1}\left\{\lambda_1 T + \sum_{n=0}^{\infty} p(n, \lambda_1 T) \max[(v_1 + n)/(\alpha_1 + T), v_2/\alpha_2]T\right\},$$

$$V_2(2) = E_{\lambda_2}\left\{\lambda_2 T + \sum_{n=0}^{\infty} p(n, \lambda_2, T) \max[v_1/\alpha_1, (v_2 + n)/(\alpha_2 + T)]T\right\}. \quad (2.8.8)$$

The difference between (2.8.5) or (2.8.7) and (2.8.8) can be phrased as follows. In (2.8.5) or (2.8.7) the fisherman's expected catch for the last period is the expectation over the true state. In (2.8.8) the fisherman's expected catch for the last period is found regardless of the true state. The philosophical dilemma is whether one should use the true value of λ, as in (2.8.7), or the expected value of λ, as in (2.8.8).

Although these value functions appear to be quite different, it will now be shown that if the Bayesian approach is used consistently there is no dilemma and that

$$\hat{V}_2(i) = V_2(i). \quad (2.8.9)$$

2.8. VALUES, STRATEGIES, AND POLICY ITERATION

It suffices to show that $\hat{V}_2(1) = V_2(1)$. To do this, we rewrite $V_2(1)$ as

$$V_2(1) = E_{\lambda_1}\left\{\lambda_1 T + \sum_{n=0}^{n_1^*} p(n, \lambda_1, T)\frac{v_2}{\alpha_2} + \sum_{n=n_1^*+1}^{\infty} p(n, \lambda_1, T)\frac{v_1+n}{\alpha_1+T}\right\}. \quad (2.8.10)$$

Since $E_{\lambda_i}\{\lambda_i\} = v_i/\alpha_i$, the first two terms on the right-hand side of (2.8.10) and (2.8.5) are the same. Consequently, $V_2(1) = \hat{V}_2(1)$ if it is true that

$$E_{\lambda_1}\left\{\sum_{n=n_1^*+1}^{\infty} p(n, \lambda_1, T)\left(\frac{v_1+n}{\alpha_1+T}\right)T\right\} = E_{\lambda_1\lambda_2}\left\{\sum_{n=n_1^*+1}^{\infty} p(n, \lambda_1, T)\lambda_1 T\right\}. \quad (2.8.11)$$

First observe that λ_2 can be eliminated from (2.8.11), since the expectation over λ_2 gives 1. Second, observe that (2.8.11) must hold termwise, that is, that (2.8.11) is true if

$$E_{\lambda_1}\{p(n, \lambda_1, T)(v_1+n)/(\alpha_1+T)\} = E_{\lambda_1}\{p(n, \lambda_1, T)\lambda_1\}. \quad (2.8.12)$$

Third, observe that $(v_1+n)/(\alpha_1+T)$ is independent of λ_1 and that

$$(v_1+n)/(\alpha_1+T)$$
$$= E\{\lambda_1 | n \text{ schools of fish found in } (0, T)\} = E\{\lambda_1 | n\}. \quad (2.8.13)$$

That is, $(v_1+n)/(\alpha_1+T)$ is the posterior expectation of λ_1, given that n schools of fish were found in $(0, t)$. Substituting (2.8.13) in (2.8.12) shows that one must verify the formula

$$E_{\lambda_1}\{p(n, \lambda_1, T)\} E\{\lambda_1 | n\} = E_{\lambda_1}\{p(n, \lambda_1, T)\lambda_1\}. \quad (2.8.14)$$

Now observe that (2.8.14) is a restatement of the definition of conditional probability, that is, that

$$E\{\lambda_1 | n\} = \frac{E\{\lambda_1 \text{ and } n \text{ schools of fish found}\}}{E\{n \text{ schools of fish found}\}}. \quad (2.8.15)$$

Consequently, (2.8.14) holds and $\hat{V}_2(i) = V_2(i)$.

Exercise 2.8.2

Redo the above calculation for an arbitrary return function $r(\lambda_i)$ as well as an arbitrary density on λ_i.

Exercise 2.8.3

Is it always true that the two approaches give the same result? What if the parameters are estimated in a non-Bayesian fashion?

Exercise 2.8.4

Another procedure for the computation of $\hat{V}_2(i)$ is the following. Let $\hat{\lambda}_i$ denote the current value of the best estimate of λ_i. For example, at the start of fishing $\hat{\lambda}_i = v_i/\alpha_i$: if n schools of fish are found in $(0, T)$, then $\hat{\lambda}_i$ is replaced by $\hat{\lambda}'_i = (v_i + n)/(\alpha_i + T)$. The procedure for the computation of $V_2(1)$ is to replace λ_i in equation (2.8.5) by the appropriate $\hat{\lambda}_i$ or $\hat{\lambda}'_i$. Then one does not have to take the expectation over λ_i and (2.8.5) is replaced by

$$\hat{V}_2(1) = \hat{\lambda}_1 T + \sum_{n=0}^{n_1^*} p(n, \hat{\lambda}_1, T)\hat{\lambda}_2 T + \sum_{n=n_1^*+1}^{\infty} p(n, \hat{\lambda}_1, T)\hat{\lambda}'_1 T$$

$$= \frac{v_1}{\alpha_1} T + \sum_{n=0}^{n_1^*} p\left(n, \frac{v_1}{\alpha_1}, T\right) \frac{v_2}{\alpha_2} T + \sum_{n=n_1^*+1}^{\infty} p\left(n, \frac{v_1}{\alpha_1}, T\right)\left(\frac{v_1 + n}{\alpha_1 + T}\right) T.$$

Note that these equations are much simpler than (2.8.5). Discuss the validity of this procedure.

Philosophical considerations aside, the idea of separating strategies and values is quite useful when one considers problems with many periods. The first useful concept is called *policy iteration*. Very often, one is interested in stationary strategies, that is, strategies that are constant in time. For continuous time problems, a stationary policy is simply

$$\pi_0 = u_0 \qquad \text{for all} \quad t. \tag{2.8.16}$$

For discrete time problems, a stationary policy is the one with the same control in each period:

$$\pi_0 = (u_0, u_0, \ldots, u_0). \tag{2.8.17}$$

The policy iteration procedure proceeds as follows.

(1) Guess an allowable stationary policy π_0.
(2) Guess a second allowable stationary policy π_1.
(3) Compute the values associated with each policy, V_1 and V_2.
(4) Replace π_0 by π_1 and π_1 by a new admissable policy, π_2 being obtained by iterating the current policy in the direction that optimizes the value function.
(5) Either return to step (3) or exit (if V_1 and V_2 are close enough to the extreme).

The numerical methods needed to utilize this sort of algorithm are described in Chapter 8.

Bibliographic Notes

The book by Bertsekas (1976) contains a good discussion of policy iteration and some numerical methods. The paper by Norman and White (1968) describes an heuristic for policy improvement procedures in which probability distributions are replaced by expectations. The book by Sengupta (1982) touches on many of the same problems treated in this chapter. The philosophical problems associated with stochastic DPEs were explicitly raised by Walters (1981), who discusses dynamic programming formulations. Engineers have already tackled such problems [Bryson and Ho (1975) or Bar Shalom and Tse (1976)].

3

Key Results in the Deterministic Theory of Resource Exploitation

The analytical machinery needed to study decision and control in uncertain resource systems is now in place. As a review, this chapter contains some key results from the deterministic theory of resource management. This chapter is in the spirit of the books by Clark (1976) and Dasgupta and Heal (1979).

3.1 CLASSIFICATION OF RESOURCE PROBLEMS AND THE CONCEPT OF UTILITY

In the conventional view resources are classified as either renewable or exhaustible. Typically, fish populations and forests fall into the first category and mineral and oil deposits fall into the second category.

A second aspect of resource problems, which is often overlooked, is that the search (or exploration) for and assessment of a resource stock is very different from the exploitation of the resource. Hence, in this book problems of optimal exploration or assessment and problems of optimal exploitation are both considered. Naturally, there is some overlap between the two, but it is useful to separate them.

Returning to the first aspect, one may ask what makes a resource renewable or not? As Dasgupta and Heal (1979) point out, often it is the time scale of the process under consideration. For example, viewed on a time

3.1. CLASSIFICATION OF RESOURCE PROBLEMS

scale of hundreds of millions of years, oil is a "renewable" resource. Similarly, viewed on a time scale of tens of years, the redwood forests in Northern California are essentially an exhaustible resource.

Exercise 3.1.1

Consider a deterministic population that grows according to the logistic equation

$$dX/dt = rX(1 - X/K). \tag{3.1.1}$$

Assume that the population has been reduced to a fraction $\alpha K (0 < \alpha < 1)$ of its carrying capacity and a harvest moratarium is imposed. Then the initial condition when solving (3.1.1) is $X(0) = \alpha K$. Find the times required for the population to reach 25%, 50%, 75%, and 95% of its carrying capacity.

The following tabulation shows some estimates for r (Clark, 1976):

Species	r
Halibut	0.71
Fin whale	0.08
Yellowfin tuna	2.61

What do the results suggest?

Perhaps the major difference between renewable and exhaustible resources is that the former offer the opportunity for sustained yields over long periods of time, if managed properly. Consequently, management involves not only the optimal exploitation rate, but one must consider the preservation and renewal of the resource.

Exercise 3.1.2

Even the point of sustained and renewed yields over time is not always perfectly clear, as this exercise will show. Consider a stock subject to the following dynamics (with X_0 given):

$$X_{n+1} = (1 - q_n)X_n e^{-M} + R((1 - q_n)X_n), \tag{3.1.2}$$

where X_n is the stock level in year n, q_n the fraction of the stock ($0 \leq q_n \leq 1$) removed in year n, M is a mortality factor, and $R(S)$ a recruitment function defined by

$$R(S) = \begin{cases} R_0 & \text{if } S > X_c, \\ R_0 S/X_c & \text{if } 0 \leq S \leq X_c. \end{cases} \tag{3.1.3}$$

This model can be used to study certain kinds of fish populations (Clark et al., 1984). As the objective function we take

$$J = \sum_{j=1}^{T} \alpha^j q_n X_n, \qquad (3.1.4)$$

where T is the time horizon and α a discount factor. Consider the problem of determining the optimal stationary harvest policy, so that q_n is a constant q for all n. Compute J as a function of q, X_0, and R_0. In particular, study the behavior of the value of q that maximizes J as a function of X_0, X_c, and R_0. Are there conditions under which the stock should be driven to extinction?

It is also worth noting that in some cases *mixed* resource systems are of interest. Such a case is studied in Chapter 7 for agricultural pest control. Basically, the problem there is that when improving yield from an agricultural crop (a renewable resource) by spraying pests with pesticide, one is trying to balance the long-run economic yield from the crop with the loss of susceptibility to the pesticide (essentially an exhaustible resource). Once the pest is resistant to pesticide, the crop cannot be protected with that pesticide. The question is, then, How does one optimally exhaust the susceptibility to pesticide?

Throughout later sections the concept of a utility function is used, and so we shall review it here. The utility function provides a measure of preference for a level of consumption, goods, or profit. For example, if $P(Q)$ is the price of a good when Q is the consumption level of the good, the social utility associated with level Q is often defined by

$$u(Q) = \int_0^Q P(q)\,dq. \qquad (3.1.5)$$

With this definition the price $P(Q) = u'(Q)$ is the marginal utility.

In general, one envisions a variable X, considered for the time being as one-dimensional, and a function $u(X)$ that characterizes preferences for levels of X. There are established ways of determining $u(X)$ in a point-wise fashion. Good discussions are found in DeGroot (1970) and Anderson, Dillon, and Hardaker (1977). An excellent example is provided in the report of Conklin et al. (1977), which estimated the utility function of expected net return over cash expenses for eight fruit orchard owners in Oregon. They found good fits with the form

$$u(X) = \alpha_1 X + \alpha_2 X^2 + \alpha_3 X^3, \qquad (3.1.6)$$

where α_1, α_2, and α_3 are fixed constants and X the expected level of return.

3.1. CLASSIFICATION OF RESOURCE PROBLEMS

Hence forth, it is assumed that $u(X)$ is known. Then the effort is oriented towards understanding how uncertainty affects the nature of preferences. One question asks, How is the utility function related to the individual's attitude towards risk?

Pratt (1964) has shown how the utility function can be related to a measure of preference for risk. To start, fix the current level X and let $\tilde{X}_1$ be a random increment in X. Define a function $p(X; X_1)$ by

$$u(X + E(\tilde{X}_1) - p(X; \tilde{X}_1)) = E\{u(X + \tilde{X}_1)\}. \tag{3.1.7}$$

With this definition $p(X; \tilde{X}_1)$ is called the *risk premium* and measures the premium needed so that the individual is indifferent between a risk of $\tilde{X}_1$ and a nonrandom amount $E(\tilde{X}_1) - p(X; \tilde{X}_1)$. That is, $p(X; \tilde{X}_1)$ is the amount that the certain level $X + E\{\tilde{X}_1\}$ must be reduced to make the individual indifferent between the certain and uncertain outcomes. Observe that $p(X; \tilde{X}_1)$ depends on $\tilde{X}_1$ only parametrically through its distribution. It is assumed that the expectations in (3.1.7) exist.

Exercise 3.1.3

Consider a gamble that pays nothing with probability 0.5 and $10 with probability 0.5 and the certain event of receiving $5. The reader should compute his or her own risk premium for such a case. Then redo the calculation for $5,000 and $10,000.

A risk $\tilde{X}_1$ is actuarially neutral if $E(\tilde{X}_1) = 0$. Consider such a risk with a small variance ε. Then one expects that the risk premium will also be small, so that $p(X, \tilde{X}_1)$ in (3.1.7) is replaced by $\varepsilon p(X, \tilde{X}_1)$. Taylor-expanding the left-hand side of (3.1.7) gives

$$u(X - \varepsilon p(X; \tilde{X}_1)) = u(X) - \varepsilon u'(X) p(X; \tilde{X}_1) + o(\varepsilon p(X; \tilde{X}_1)). \tag{3.1.8}$$

Taylor-expanding the right-hand side of (3.1.7) gives

$$E\{u(X + \tilde{X}_1)\} = E\{u(X) + u'(X)\tilde{X}_1 + \tfrac{1}{2}u''(X)\tilde{X}_1^2 + \cdots\}$$
$$= u(X) + \tfrac{1}{2}\varepsilon u''(X) + o(\varepsilon). \tag{3.1.9}$$

Setting (3.1.8) and (3.1.9) equal gives, to leading order in $p\varepsilon$,

$$p(X; \tilde{X}_1) = -\tfrac{1}{2}[u''(X)/u'(X)]. \tag{3.1.10}$$

Pratt defines the measure of local risk aversion $r(X)$ by

$$r(X) \equiv -u''(X)/u'(X), \tag{3.1.11}$$

since $-r(X)$ is a local measure for liking risk or having a propensity to gamble. Equation (3.1.11) allows one to interpret risk aversion in terms of the concavity of the utility function. It seems reasonable to assume that

$u'(X) > 0$. Then $r(X) > 0$ if $u''(X) < 0$, which means that the utility function $u(X)$ is concave downward. A quick sketch of such a curve shows that a certain outcome is always preferred to an uncertain outcome with the same expectation. For example, for the utility function (3.1.6), $r(X)$ is given by

$$r(X) = -\frac{2\alpha_2 + 6\alpha_3 X}{\alpha_1 + 2\alpha_2 X + 3\alpha_2 X^2}. \tag{3.1.12}$$

Exercise 3.1.4

Describe $r(X)$ in (3.1.12) as a function of X, $X \geq 0$, for all possible sets of parameters (i.e., the α_i may be positive, negative, or zero).

Common assumptions are the cases of constant risk aversion and decreasing risk aversion. If $r(X)$ is a constant γ, then $u(X)$ is determined by solving the differential equation

$$u''(X) = -\gamma u'(X). \tag{3.1.13}$$

Exercise 3.1.5

Solve (3.1.13) for all possible cases (i.e., all possible values of γ).

Pratt (1964) provides examples of utility functions for which $r(X)$ is strictly decreasing.

Exercise 3.1.6

Suppose that X is a vector with components $X_1, X_2, \ldots, X_n$. How should these calculations be modified?

Pratt (1964) introduces a new function $r^*(X)$ defined by

$$r^*(X) = Xr(X) \tag{3.1.14}$$

and calls it the local proportional risk aversion function at point X.

Exercise 3.1.7

What kind of utility functions give constant local proportional risk aversion functions?

Pratt (1964) also mentions that there appear to be cases in which $r^*(X)$ is first decreasing and then increasing. A model for this kind of function is

$$r^*(X) = \alpha_0 + \alpha_1 X + \alpha_2 X^2, \tag{3.1.15}$$

3.2 EXPLOITATION OF AN EXHAUSTIBLE RESOURCE

with $\alpha_1 < 0$ and $\alpha_2 > 0$. If $r^*(X)$ is given by (3.1.15), then $u(X)$ satisfies the differential equation

$$u''(X) = -u'(X)(\alpha_0/X + \alpha_1 + \alpha_2 X). \tag{3.1.16}$$

Integrating (3.1.16) once gives

$$\ln[u'(X)] = -(\alpha_0 \ln X + \alpha_1 X + \alpha_2 X^2/2) + c_0, \tag{3.1.17}$$

where c_0 is an integration constant. From (3.1.17) one finds

$$u(X) = c_1 \int_{X_0}^{X} s^{-\alpha_0} \exp[-\alpha_1 s - \alpha_2 s^2/2] \, ds + c_2, \tag{3.1.18}$$

where c_1 and c_2 are again constants resulting from the integration and X_0 is a reference level chosen so that $u(X_0) = 0$.

Exercise 3.1.8

Describe the behavior of $u(X)$ given by (3.1.18) as a function of the parameters α_0, α_1, α_2, and X_0.

Exercise 3.1.9

Assume that X is normally distributed with mean μ and variance σ^2. Find an approximation for $E\{u(X)\}$ if $u(X)$ is given by (3.1.18).

Bibliographic Notes

The books of DeGroot (1970) and Anderson, Dillon, and Hardaker (1977) contain good discussions of utility and how it can be measured. Pratt's excellent paper is reprinted in the book edited by Diamond and Rothschild (1978), which contains many other interesting articles as well. Schoemaker (1982) provides a good review of the expected utility model and its support and limitations. There are a number of cases in which the expected utility model fails. Kahneman and Tversky (1979) developed a new theory to account for such cases, called the "prospect theory."

3.2 EXPLOITATION OF AND EXHAUSTIBLE RESOURCE: OPTIMAL EXTRACTION POLICY AND PRICE DYNAMICS

In this section and the next, the deterministic theory of resource extraction is briefly reviewed. These reviews are not intended to be comprehensive; rather the goal is to remind the reader of some of the main points

3. THE DETERMINISTIC THEORY OF RESOURCE EXPLOITATION

in the deterministic theory of resource exploitation. This helps fix notation and guides the questions that one can ask in the stochastic case.

This section is concerned with the optimal exploitation of an exhaustible resource. Consider a single resource stock $X(t)$ known with certainty at time 0, that is, $X(0) = X_0$ is given. Assume that the price per unit of resource is $p(t)$, the cost of extraction when the stock level is x is $c(x)$, the discount rate is r, and the social utility of extraction is $u(q, p, c)$, where q is the extraction rate. Consider the problem of maximizing the discounted stream of social utility over an infinite time horizon. This gives an optimization problem

$$\max_q \int_0^\infty e^{-rt} u(q(t), p(t), c(X(t))) \, dt, \qquad (3.2.1)$$

such that $q \geq 0$, $dX/dt = -q$, $X(0) = X_0$. This problem can be solved using either the calculus of variations or optimal control theory. Both theories are reviewed in the appendix to this chapter. As a review, problem (3.2.1) will now be solved using both methods.

Since the total amount of resource is constant, $q(t)$ satisfies the condition

$$\int_0^\infty q(t) \, dt = X_0. \qquad (3.2.2)$$

[To obtain (3.2.2) one assumes that all the resource is extracted, of course. When might it not be?] Constraint (3.2.2) is adjoined to problem (3.2.1) by means of a Lagrange multiplier, giving the unconstrained problem

$$\max_{q \geq 0} \int_0^\infty [e^{-rt} u(q(t), p(t), c(X(t))) - \lambda q] \, dt - \lambda X_0, \qquad (3.2.3)$$

where $p(t)$ is exogenous and presumed known, $\dot{X} = -q$, and λ is the Lagrange multiplier. Set

$$L(t, X, q) = e^{-rt} u(q(t), p(t), c(X(t))) - \lambda q. \qquad (3.2.4)$$

In this case the Euler equation of the calculus of variations [Courant and Hilber (1953), reviewed in the appendix of this chapter] becomes (recall that $\dot{X} = -q$)

$$\partial L/\partial X = -d(\partial L/\partial q)/dt. \qquad (3.2.5)$$

That is, (3.2.5) is satisfied on the optimal extraction path. Substituting (3.2.4) in (3.2.5) gives

$$e^{-rt} \frac{\partial u}{\partial c} c'(X(t)) = -\frac{d}{dt}\left[e^{-rt} \frac{\partial u}{\partial q} - \lambda\right]. \qquad (3.2.6)$$

3.2 EXPLOITATION OF AN EXHAUSTIBLE RESOURCE

Since $\partial u/\partial q$ is a price-like variable, (3.2.6) shows that λ is also a price-like variable.

In the simplest case $c'(x) = 0$ and $\partial u/\partial q$ is replaced by $p(t)$. Equation (3.2.6) becomes

$$p(t) = \lambda e^{rt}. \tag{3.2.7}$$

Equation (3.2.7) shows that the price rises at the social rate of discount on the optimal path. This is the classic result in exhaustible resource economics [see Hotelling (1931)].

What is the optimal consumption path? If $c'(x) = 0$, the general Euler equation becomes

$$\partial u(q, p, c)/\partial q = (\lambda + k_1)e^{rt} \equiv \hat{\lambda}e^{rt}, \tag{3.2.8}$$

where k_1 is a constant of integration.

It may occur that (3.2.8) can be solved for $q(t)$, in which case one formally writes

$$q(t) = u_q^{-1}(\hat{\lambda}e^{rt}, p, c), \tag{3.2.9}$$

where u_q^{-1} is the inverse function to $\partial u/\partial q$.

One common form for $u(q, p, c)$ is simply profit, given by (with the time dependence of $X(t)$ suppressed for convenience)

$$u(q(t), p(t), c(X)) = (p(t) - c(X))q(t). \tag{3.2.10}$$

In this case the Euler equation is

$$e^{-rt}c'(X)q(t) = d[e^{-rt}(p - c(X))]/dt. \tag{3.2.11}$$

If $c'(X) \neq 0$, then (3.2.11) can be solved for $q(t)$ if $p(t)$ is given.

Exercise 3.2.1

A common model for costs is $c(x) = c_0/X$. What heuristic motivation can you give for this choice? Solve (3.2.11) using $c(x) = c_0/X$.

Equation (3.2.11) can be used to study price dynamics. Differentiating the right-hand side of (3.2.11) and rearranging gives

$$d(p - c(X))/dt = r(p - c(X)) + c'(X)q(t). \tag{3.2.12}$$

Dividing both sides by $p - c(X)$ gives

$$\frac{d(p - c(X))/dt}{p - c(X)} = r + \frac{c'(X)q(t)}{p - c(X)}. \tag{3.2.13}$$

If $c'(X) = 0$, then (3.2.13) reduces to a result analogous to the condition (3.2.7).

In general, one expects that $c'(X) < 0$ (the cost of extraction at a stock level X decreases as the stock level increases). The second term on the right-hand side of (3.2.13) can be viewed as an effective discount rate composed of the true discount rate and a perturbation of it.

Exercise 3.2.2

Is the extraction policy with $c'(X) \neq 0$ more or less conservative than the policy with $c'(X) = 0$?

Evaluating the derivative in (3.2.13) and using the fact that $dc(X)/dt = -c'(X)q$ (since $\dot{X} = -q$) gives

$$dp/dt = r(p - c(X)). \tag{3.2.14}$$

Equation (3.2.14) provides a way of characterizing the price path explicitly.

We shall now solve this same problem using optimal control theory (see the appendix to this chapter) for the case when the utility function is profit. The mathematical formulation is essentially the same as (3.2.1):

$$\max_q \int_0^\infty e^{-rt}[p - c(X)]q\, dt \quad \text{such that} \quad \dot{X} = -q,\ X(0) = X_0,\ q \geq 0. \tag{3.2.15}$$

The Hamiltonian for this problem is

$$\mathcal{H} = e^{-rt}(p - c(X))q - \lambda q. \tag{3.2.16}$$

Observe that the Hamiltonian (3.2.16) is essentially the integrand in (3.2.3) [for the utility function given by (3.2.10)], which is obtained by adjoining the Lagrange multiplier to (3.2.1). Observe also that λ is still a price-like variable; it is called the *shadow price* of the resource. The dynamics of λ are given by

$$\dot{\lambda} = -\partial \mathcal{H}/\partial X = e^{-rt}c'(X)q. \tag{3.2.17}$$

According to the maximum principle, the optimal extraction rate q^* maximizes $\mathcal{H}$. In this case, when the utility function is linear in $q(t)$, the control is "bang-bang," that is, if

$$e^{-rt}(p - c(X)) < \lambda, \tag{3.2.18}$$

then $q(t)$ should be as small as possible. Typically, this will mean that $q(t) = 0$.

If

$$e^{-rt}(p - c(X)) > \lambda, \tag{3.2.19}$$

3.2 EXPLOITATION OF AN EXHAUSTIBLE RESOURCE

then $q(t)$ should be as large as possible. Typically, this will mean that $q(t) = q_{\max}$, the upper limit on the extraction rate.

The third possibility is that

$$e^{-rt}(p - c(X)) = \lambda. \tag{3.2.20}$$

In this case, the Hamiltonian $\mathscr{H}$ is independent of $q(t)$. This case is called the *singular* case and it is said that the *singular path* characterizes $q(t)$. Differentiating (3.2.20) gives

$$-re^{-rt}(p - c(X)) + e^{-rt}\frac{d}{dt}(p - c(X)) = \frac{d\lambda}{dt} \tag{3.2.21}$$

and substituting (3.2.17) in (3.2.21) gives (3.2.12), so that the two methods give completely equivalent results.

As a kind of "warm-up problem" for the next chapter, we now show how to modify the problem just solved for the situation when an uncertain tax on profits exists. Assume that a tax τ_p exists on profits, reducing the profits to $1 - \tau_p$ of their original value. The tax is uncertain in the sense that there is a time $\tilde{t}_1$ such that

$$\begin{aligned} \tau_p > 0 &\quad \text{for} \quad 0 < t < \tilde{t}_1, \\ \tau_p = 0 &\quad \text{for} \quad \tilde{t}_1 \le t \le \infty. \end{aligned} \tag{3.2.22}$$

Here $\tilde{t}_1$ is a random variable with density $f(t_1)$. Problem (3.2.1) is now replaced by

$$\max_q \mathrm{E}_{\tilde{t}_1}\left\{\int_0^\infty e^{-rt}[p(t) - c(X(t))]q(t)(1 - \tau_p)\,dt\right\},$$

such that $dX/dt = -q$, $X(0) = X_0$, $q \ge 0$, \quad(3.2.23)

where $\mathrm{E}_{\tilde{t}_1}$ is the expectation over the density for $\tilde{t}_1$. The problem posed by (3.2.23) is indeed formidable, but the way that uncertainty enters the problem allows one to make some progress. Observe that the integrand can be rewritten as

$$\begin{aligned}\max_q \mathrm{E}_{\tilde{t}_1}\bigg\{&\int_0^{\tilde{t}_1} e^{-rt}[p(t) - c(X(t))]q(t)(1 - \tau_p)\,dt \\ &+ \int_{\tilde{t}_1}^\infty e^{-rt}[p(t) - c(X(t))]q(t)\,dt\bigg\}.\end{aligned} \tag{3.2.24}$$

Taking the expectation in (3.2.24) gives

$$\begin{aligned}\max_q \int_0^\infty f(t_1)\bigg\{&\int_0^{t_1} e^{-rt}[p(t) - c(X(t))]q(t)(1 - \tau_p)\,dt \\ &+ \int_{t_1}^\infty e^{-rt}[p(t) - c(X(t))]q(t)\,dt\bigg\}\,dt_1.\end{aligned} \tag{3.2.25}$$

84   3. THE DETERMINISTIC THEORY OF RESOURCE EXPLOITATION

Observe that the stochastic problem has been reduced to a more complicated deterministic problem. To see how to proceed further, set

$$J^*[X_1, t_1] = \max_q \int_{t_1}^{\infty} e^{-rt}[p(t) - c(X(t))]q(t)\,dt, \qquad q \geq 0, \quad (3.2.26)$$

such that $X(t_1) = X_1$. The interpretation of $J^*[X_1, t]$ is that it is the optimal value of the extraction policy, given that the tax switched at time t_1 with the resource level at X_1. Next, let $F(t_1)$ be the distribution function corresponding to $f(t_1)$. The first term in (3.2.25) can be integrated by parts over t_1 as follows. First, for simplicity set

$$R(t) = e^{-rt}[p(t) - c(X(t))]q(t)(1 - \tau_p). \qquad (3.2.27)$$

Then the integration by parts is given by

$$\int_0^{\infty} f(t_1)\left[\int_0^{t_1} R(t)\,dt\right]dt_1 = -[1 - F(t_1)]\int_0^{t_1} R(t)\,dt \Big|_{t_1=0}^{t_1=\infty}$$

$$+ \int_0^{\infty} [1 - F(t_1)]R(t_1)\,dt_1. \qquad (3.2.28)$$

Assuming that $F(\infty) = 1$, the first term on the right-hand side of (3.2.28) vanishes.

Exercise 3.2.3

Interpret the second integral in the right-hand side of (2.2.28). [*Hint*: $F(t_1) = \Pr\{\tilde{t}_1 \leq t_1\}$ is the probability that the tax is in force at t_1.]

Substituting (3.2.28) in (3.2.25) and then in (3.2.23), the problem of interest is now

$$\max_q \int_0^{\infty} \{[1 - F(t_1)]e^{-rt}[p(t) - c(X(t))]q(1 - \tau_p) + f(t_1)J^*[X(t_1)t]\}\,dt_1,$$

such that $dX/dt = -q$, $X(0) = X_0$, $q \geq 0$. (3.2.29)

The problem posed in (3.2.29) is now a deterministic control problem.

Exercise 3.2.4

Solve (3.2.29) using optimal control theory.

The reason that the original stochastic problem could be converted to a deterministic one is that the uncertainty entered in a very simple way. In general, one is not so lucky.

Bibliographic Notes

Clark (1976, pp. 135–152) and Pindyck (1980) give more thorough discussions of the optimal exploitation of a known exhaustible resource stock. Clark describes optimal control theory as well. Other good references for control theory and the calculus of variations are the papers by Dreyfus (1972) and Dorfman (1969) and the book by Kamien and Schwartz (1981). The problem of the uncertain tax is described by Stefanou (1981), in which a complete description of the optimal path is given.

3.3 EXPLOITATION OF A RENEWABLE RESOURCE; OPTIMAL HARVEST RATES AND PRICE DYNAMICS

We now assume that the stock dynamics satisfy

$$\dot{x} = f(x) - q(t), \tag{3.3.1}$$

where $f(x)$ is a given growth function and $q(t)$ the extraction rate. For example, $f(x)$ could be the logistic function

$$f(x) = rx(1 - x/K)$$

and $q(t)$ could be given by

$$q(t) = q_0 E(t) x(t),$$

where q_0 and $E(t)$ are parameters. In this case (3.3.1) is often called the Schaefer model (Schaefer, 1957).

For simplicity, it is assumed in this section that the utility function is simply net profit, given by $[p(t) - c(x)]q(t)$, where $p(t)$ and $c(x)$ have the same interpretations as in the previous section. The optimization problem is then given by

$$\max_q \int_0^\infty e^{-rt}[p(t) - c(x)]q(t)\,dt \quad \text{such that}$$

$$\dot{x} = f(x) - q(t), \quad x(0) = x_0, \quad q \geq 0. \tag{3.3.2}$$

To apply the calculus of variations, observe that $q(t) = f(x) - \dot{x}$. Then the functional in (3.3.2) becomes

$$\max_q \int_0^\infty e^{-rt}[p(t) - c(x)][f(x) - \dot{x}]\,dt. \tag{3.3.3}$$

In the rest of this section, the argument of $p(t)$ will be suppressed. The Euler equation of the calculus of variations for this problem is

$$-\frac{d}{dt}\{e^{-rt}[p - c(x)]\} = e^{-rt}\frac{\partial}{\partial x}\{[p - c(x)][f(x) - \dot{x}]\}. \quad (3.3.4)$$

There are a number of points of departure from (3.3.4). The first involves the assumption that p is constant. In this case, taking all derivatives in (3.3.4) and solving the resulting equation gives

$$f'(x) = r + \frac{c'(x)f(x)}{p - c(x)}. \quad (3.3.5)$$

Equation (3.3.5) specifies an optimal equilibrium level for the stock [see Clark (1976) for a fuller discussion of (3.3.5)].

Equation (3.3.4) can also be used to determine the dynamics of $p - c(x)$ and p itself in general. To do this, rearrange (3.3.4) as follows:

$$r[p - c(x)] - d[p - c(x)]/dt = -c'(x)[f(x) - \dot{x}] + f'(x)[p - c(x)]. \quad (3.3.6)$$

Equation (3.3.6) is an equation for the dynamics of the unit net revenue. Since $f(x) - \dot{x} = q$, it becomes

$$d[p - c(x)]/dt = r[p - c(x)] + c'(x)q - f'(x)[p - c(x)]. \quad (3.3.7)$$

Equation (3.3.7) can be rewritten to give a generalization of the condition (3.2.7) as applied to renewable resources. Dividing both sides of (3.3.6) by $p - c(x)$ gives

$$\frac{1}{p - c(x)}\frac{d}{dt}[p - c(x)] = r - f'(x) + \frac{c'(x)}{p - c(x)}. \quad (3.3.8)$$

Typically, $f(x)$ will be similar to a logistic curve (Fig. 3.3.1), so that $f'(x)$ may be positive, zero, or negative.

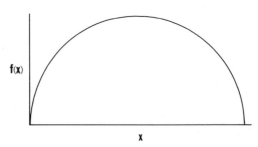

Fig. 3.3.1 The stock growth function $f(x)$ is often similar to a logistic curve. This means that $f'(x)$ is positive, zero, or negative depending upon the value of x.

Exercise 3.3.1

Assume that $c'(x) \equiv 0$. Discuss the effective rate of discount $r - f'(x)$ using Fig. 3.3.1.

The dynamics of the price can also be obtained from (3.3.7), since $\dot{x} = f(x) - q$. Evaluating the derivative $dc(x(t))/dt$ gives

$$dp/dt = r[p - c(x)] - f'(x)[p - c(x)] + c'(x)f(x). \qquad (3.3.9)$$

In this case it makes sense to speak of a stock-price equilibrium. That is, in the (x, p) plane (3.3.1) and (3.3.9) describe the dynamics of the variables x and p. One can ask if there is a point (presumably for which both x and p are positive) at which the right-hand sides of (3.3.1) and (3.3.9) are zero. This would then be an equilibrium point since at this point $dx/dt = dp/dt = 0$. Setting $\dot{x} = 0$ gives

$$f(x) = q. \qquad (3.3.10)$$

Setting $dp/dt = 0$ gives

$$p_{eq} = c(x) - [c'(x)f(x)]/[r - f'(x)]. \qquad (3.3.11)$$

Exercise 3.3.2

Interpret (3.3.11) and characterize the equilibrium in the (x, p) phase plane. How are these results related to (3.3.5)?

Exercise 3.3.3

Rederive the results of this section using optimal control theory.

Bibliographic Notes

Again, Clark (1976) and Pindyck (1983) contain more thorough discussions of the optimal management of renewable resource stocks. The literature on such problems is now vast. Clark (1976), Dasgupta and Heal (1979), and Pindyck (1983) provide good sources for other references.

APPENDIX A: A REVIEW OF THE CALCULUS OF VARIATIONS AND OPTIMAL CONTROL THEORY

This appendix contains a review of the simplest problem of the calculus of variations and optimal control theory as well as the conclusion of Example 1.5.4.

3. THE DETERMINISTIC THEORY OF RESOURCE EXPLOITATION

To begin, consider the optimization problem

$$\max \int_{t_0}^{t_1} L(t, x, \dot{x}) \, dt \quad \text{such that} \quad x(t_0) = x_0, x(t_1) = x_1. \tag{3.A.1}$$

The problem here is to find a piecewise continuous path $x(t)$ that joins x_0 and x_1 and also maximizes the integral in (3.A.1).

Here $L(t, x, \dot{x})$ is viewed as some exogenously given function. For a given path $x(t)$, let $J[x]$ be defined by

$$J[x] = \int_{t_0}^{t_1} L(t, x, \dot{x}) \, dx. \tag{3.A.2}$$

Thus, $J[x]$ is shorthand for $J[x(t)]$.

Suppose that x^* is the extremizing path. Then any other path can be written as

$$x(t) = x^*(t) + \varepsilon y(t), \tag{3.A.3}$$

where ε is a parameter and $y(t)$ an arbitrary piecewise continuous function subject to the following condition. For $x(t)$ to be an allowable path, it must be true that $x(t_i) = x_i$. Since $x^*(t_i) \equiv x_i$, one concludes that

$$y(t_0) = y(t_1) = 0. \tag{3.A.4}$$

We now consider the variation $\delta J[x^*, x, \varepsilon]$ defined by

$$\delta J[x^*, x, \varepsilon] = J[x^*] - J[x] \tag{3.A.5}$$

$$= \int_{t_0}^{t_1} \{L(t, x^*, \dot{x}^*) - L(t, x, \dot{x})\} \, dt. \tag{3.A.6}$$

When $\varepsilon = 0$, $x = x^*$ and $\delta J[x^*, x, 0] = 0$. For any other $\varepsilon \neq 0$, $\delta J[x^*, x, 0] \geq 0$ by assumption. Thus, as a function of ε, $\delta J[x^*, x, \varepsilon]$ has a minimum at $\varepsilon = 0$. Consequently, one can write that

$$\frac{d}{d\varepsilon} \left[\int_{t_0}^{t_1} \{L(t, x^*, \dot{x}^*) - L(t, x^* + \varepsilon y, \dot{x}^* + \varepsilon \dot{y})\} \, dt \right]\bigg|_{\varepsilon=0} = 0. \tag{3.A.7}$$

Taylor-expanding $L(t, x^* + y, \dot{x}^* + \varepsilon \dot{y})$ around $(t, x^*, \dot{x}^*)$, taking the derivative, and setting $\varepsilon = 0$ gives

$$\int_{t_0}^{t_1} \left(\frac{\partial L}{\partial x}(t, x^*, \dot{x}^*) y + \frac{\partial L}{\partial \dot{x}}(t, x^*, \dot{x}^*) \dot{y} \right) dt = 0. \tag{3.A.8}$$

APPENDIX A

Integrating the second term in (3.A.8) by parts and applying (3.A.4) gives

$$\int_{t_0}^{t_1} \left\{ \frac{\partial L}{\partial x}(t, x^*, \dot{x}^*) - \frac{d}{dt}\left(\frac{\partial L}{\partial \dot{x}}(t, x^*, \dot{x}^*)\right) \right\} y(t) \, dt = 0. \quad (3.A.9)$$

Since (3.A.9) must hold for all $y(t)$ satisfying (3.A.4), it must be true that

$$\frac{\partial L}{\partial x}(t, x^*, \dot{x}^*) = \frac{d}{dt}\left(\frac{\partial L}{\partial \dot{x}}(t, x^*, \dot{x}^*)\right). \quad (3.A.10)$$

If (3.A.10) did not hold, then one could pick a $y(t)$ that has a large spike near a nonzero point of $\partial L/\partial x = (d/dt)(\partial L/\partial \dot{x})$ and thus make the integral (3.A.9) as large as desired.

Exercise 3.A1

Equation (3.A.10) is a necessary condition. How could one find necessary and sufficient conditions?

Equation (3.A.10) is the Euler equation or Euler–Lagrange equation of the calculus of variations. The problem (3.A.1) has now been converted to the following: find $x(t)$, twice differentiable, such that

$$\frac{\partial L}{\partial x}(t, x, \dot{x}) = \frac{d}{dt}\left(\frac{\partial L}{\partial \dot{x}}(t, x, \dot{x})\right),$$

$$x(t_0) = x_0, \qquad x(t_1) = x_1. \quad (3.A.11)$$

The problem posed by (3.A.11) is a two-point boundary value problem. Its solution need not exist, nor, if it exists, be unique. Similarly, the solution of the variational problem (3.A.1) need not exist.

Exercise 3.A.2

Does the following problem have a solution?

$$\min \int_0^1 x(t) \, dt \quad \text{such that} \quad x(0) = x(1) = 1.$$

Optimal control theory applies to a more general variational problem. Here the dynamics are given by

$$dx/dt = f(t, x, u), \quad (3.A.12)$$

where $f(t, x, u)$ is a known function and $u(t)$ a control from a control set U. The optimization problem is

$$\max_{u \in U} \int_{t_0}^{t_1} L(t, x, u) \, dt + \Phi(x(t_1), t_1)$$

$$\text{such that} \quad (3.A.12) \text{ holds and } x(t_0) = x_0. \quad (3.A.13)$$

This problem is solved using an algorithm developed from the Pontryagin maximum principle (PMP). The PMP is difficult to derive rigorously [see Fleming and Rishel (1975) or Dreyfus (1972)], so the algorithm will simply be stated here.

First, form the Hamiltonian $\mathcal{H}(t, x, u, \lambda)$ given ny

$$\mathcal{H}(t, x, u, \lambda) = L(t, x, u) + \lambda(t) f(t, x, u). \tag{3.A.14}$$

Note that the dynamics of $x(t)$ are then

$$dx/dt = \partial \mathcal{H}/\partial \lambda. \tag{3.A.15}$$

According to the PMP, if $u^*(t)$ is the optimal control and $x^*(t)$ the corresponding path, then

$$\mathcal{H}(t, x^*, u^*, \lambda) = \max_{u \in U} \mathcal{H}(t, x, u, \lambda)$$

and $\lambda(t)$ satisfies the dynamical equation

$$d\lambda/dt = -\partial \mathcal{H}/\partial x. \tag{3.A.16}$$

Exercise 3.A.3

Show that (3.A.12–3.A.16) reduce to (3.A.1) and (3.A.10) if $f(t, x, u) = u$. What additional assumptions are needed?

Exercise 3.A.4

Solve problem (3.A.13) by first adjoining the constraint (3.A.12) to the integral by a Lagrange multiplier and then applying the calculus of variations. Why is this not a proof of the PMP?

Equations (3.A.12) and (3.A.16) each require an initial condition for integration. One condition is $x(t_0) = x_0$. The other condition [on $\lambda(t)$] depends on the final condition on $x(t)$. The most common conditions encountered in this book are shown in the following tabulation:

Condition on x	Condition on λ
$x(t_1)$ given	None allowed
$x(t_1)$ free	$\lambda(t_1) = \partial \Phi/\partial x$
$x(t_1) \geq 0$	$\lambda(t_1) \geq \partial \Phi/\partial x_1$
	and $x(t_1)[\lambda(t_1) - \partial \Phi/\partial x_1] = 0$

APPENDIX A

To illustrate how the PMP is used, reconsider Example 1.5.4. The problem there is [Eq. (1.5.37)]

$$\text{minimize } c_1 V(T) + c_2 \int_0^T u(s)\, ds$$

$$\text{such that } \frac{dV}{dt} = 2f(t)V(t) - \frac{V(t)^2 \beta u(t)^{n+m}}{\gamma + u(t)^m}. \quad (3.A.17)$$

In this case the Hamiltonian is

$$\mathcal{H} = c_2 u + \lambda \left\{ 2fV - \frac{V^2 \beta u^{n+m}}{\gamma + u^m} \right\}. \quad (3.A.18)$$

Assuming an internal optimum, the extreme value of u satisfies $\partial \mathcal{H}/\partial u = 0$. This gives

$$c_2 = \frac{\lambda V^2 \beta}{(\gamma + u^m)^2} \left[(n+m) u^{n+m-1} (\gamma + u^m) - m u^{n+2m-1} \right]. \quad (3.A.19)$$

Consider the special case where $m = n - 1$. Then (3.A.19) becomes

$$c_2 = \frac{\lambda V^2 \beta}{(\gamma + u)^2} [2u\gamma + u^2]. \quad (3.A.20)$$

Equation (3.A.20) can be solved to give u explicitly as a function of λ and V. The dynamics of λ are found from

$$d\lambda/dt = -\partial \mathcal{H}/\partial V \quad (3.A.21)$$

with the boundary condition

$$\lambda(T) = c_1. \quad (3.A.22)$$

From (3.A.18) the dynamics of λ are

$$\frac{d\lambda}{dt} = -\lambda \left\{ 2f - \frac{2V\beta u^{n+m}}{\gamma + u^m} \right\}. \quad (3.A.23)$$

For the special case where $n = m = 1$, the solution to the optimization problem is found by solving (3.A.20) along with

$$\frac{dV}{dt} = 2fV - \frac{V^2 \beta u^2}{\gamma + u}, \quad V(0) = V_0,$$

$$\frac{d\lambda}{dt} = -2\lambda \left[f - \frac{V\beta u^2}{\gamma + u} \right], \quad \lambda(T) = c_1. \quad (3.A.24)$$

Exercise 3.A.5

What are the hidden assumptions in the derivation of (3.A.24)?

Exercise 3.A.6

Solve (3.A.24).

Bibliographic Notes

The books by Fleming and Rishel (1975) and Kamien and Schwartz (1981) and the paper by Dreyfus (1972) provide good discussions of the topics reviewed here. Dorfman (1969) shows how many of the results of optimal control theory can be derived using economic reasoning from capital theory. Other books on the calculus of variations and optimal control theory are Hestenes (1980), Intriligator (1971), Kalaba and Springarn (1982), and Leitmann (1981). Kalaba and Springarn (1982), in particular, discuss numerical methods needed to solve optimal control problems.

4

Exploration for and Optimal Utilization of Exhaustible Resources

Problems involving exhaustible resources are slightly easier than those involving renewable resources because they lack complicated stock dynamics. (This is, of course, a generalization for which many counterexamples exist.) Consequently, we shall treat problems of exhaustible resources first.

Three papers, two contemporary and one quite old, provide a good background for this chapter. The paper by Peterson and Fisher (1977) is an excellent survey of the literature on managing "extractive resources" and provides a good historical review of the problems associated with exhaustible resources.

The paper by Allais (1957), although now somewhat out of date, is an interesting example of operations analysis in resource assessment. Allais analyzed the potential for mining in the Algerian Sahara. He begins with a model for mining exploration. This model assumes that the indicators of a deposit of minerals are given by a Poisson model for detection. The actual value of deposits is assumed to be lognormal, that is, if Z_j is the size of the jth deposit, then $\log(Z_j)$ is normally distributed. Allais next considered how the parameters in these models could be estimated. The parameters are: (1) the parameter of the Poisson process and (2) the mean and variance of the normal density for $\log(Z_j)$. The parameters were estimated by using statistics from France, North Africa, the United States, and other countries.

After estimating the parameters, Allais estimated the probability of success of exploration and the costs of exploration. From these estimates it was possible to estimate the overall profitability of mining and Sahara.

The paper by Epple and Lave (1982) on the controversy surrounding the management of helium reserves stresses the effect of discount the rate and assesses the implications of various management policies. Although exploration is not a major problem in the helium controversy, this paper is still very interesting as background material.

Much of the analysis of resource management (and this book, to a large extent, is no exception) hinges on the assumptions of a known discount rate and the objective of maximizing a discounted stream of utility. Although these are useful concepts, there is continual questioning of their applicability in the "real world." Consequently, the analyst should proceed cautiously when converting from analytical models to real world decisions.

This chapter begins with a study of the exploration and search for nonrenewable resources. Then a series of models for optimal utilization of such resources is developed. It is important to consider *exploration* as well as *exploitation* when dealing with natural resource problems. Any theory that does not consider both aspects of the problem cannot be complete.

4.1 EXPLORATION FOR EXHAUSTIBLE RESOURCES

Before a resource can be exploited, it must be found and assessed in terms of size and quality. Consequently, a natural starting point is the analysis of exploration. Underlying any analysis of exploration must be a model for the distribution of the resource. Griffiths (1966) and Uhler and Bradley (1970) studied the distribution of oil wells and derived models similar to the ones given here. It what follows the term "resource pool" refers to any kind of immobile, exhaustible resource stock.

Let N be the number of resource pools in a given region of area A. Assume that N has a Poisson distribution with parameter λ. Then, given λ,

$$\Pr\{N = k | \lambda\} = e^{-\lambda} \lambda^k / k!. \tag{4.1.1}$$

Very often λ is unknown too, so that one must associate a distribution with it. Thus, assume that λ has a gamma distribution with parameters v, α. This leads to the following negative binomial distribution for N:

$$\Pr\{N = k\} = \frac{1}{k!} \frac{\alpha^v}{(\alpha + 1)^{v+k}} \frac{\Gamma(v + k)}{\Gamma(v)}. \tag{4.1.2}$$

Table 4.4.1

Comparison of Poisson and Negative Binomial Models for the Distribution of Oil Deposits in 5 mi × 5 mi Cells[a]

Deposits	Observed	Frequency	
		Negative binomial	Poisson
0	8586	8584.3	8508.53
1	176	176.8	303.01
2	35	39.1	5.4
3	13	11.3	0.06
4	6	3.6	0
5	1	1.2	0
6	0	0.4	0
7	0	0.2	0
8	0	0.06	0
9	0	0.02	0
≥ 10	0	0.01	0
χ^2	—	1.086	6365.91

[a] From Uhler and Bradley (1970).

Uhler and Bradley compared the fits of (4.1.1) and (4.1.2) with data on oil deposits in 5 mi × 5 mi grids from Alberta, Canada. Their results are shown in Table 4.4.1. The negative binomial distribution provides a much better fit to the data than the Poisson distribution. This should not be interpreted, however, to mean that the negative binomial distribution is "right" and the Poisson distribution is "wrong." In general, one should expect that any two-parameter distribution will fit a set of data better than any one-parameter distribution. The excellent fit of the negative binomial model is encouraging. Although that model is derived for mathematical convenience, it appears to work reasonably well.

In the ecological literature [see, e.g., Pielou (1977)], the negative binomial distribution is associated with "contagion" or "clumping." On physical grounds, it seems quite reasonable to assume a clumped distribution for oil deposits. The results of Uhler and Bradley certainly support such an assumption.

Once discovered, the size of a deposit must be assessed. It is likely that different deposits will have different sizes. Let S_i be the size of the ith deposit. Since many factors contribute to the determination of the size of a deposit, it may be reasonable to assume that $\log S_i$ is normally distributed with mean μ and variance σ^2.

Exercise 4.1.1

Criticize the assumptions introduced thus far and suggest alternate hypotheses.

The lognormal assumption means that

$$\Pr\{\log S_i \in (s, s+ds)\} = \frac{1}{\sqrt{2\pi}\,\sigma} \exp[-(s-\mu)^2/2\sigma^2]\,ds. \quad (4.1.3)$$

The total deposit in the region, $Z(N)$, is given by

$$Z(N) = \sum_{i=1}^{N} S_i. \quad (4.1.4)$$

Recall that both N and S_i are random variables. The moments of $Z(N)$ are easily found using the result of Chapter 1. In particular, if Z and N are independent (which is a reasonable first approximation), then these moments are

$$\begin{aligned} E\{Z\} &= E\{N\}\,E\{S_i\}, \\ \mathrm{Var}\{Z\} &= E\{S_i\}^2\,\mathrm{Var}\{N\} + E\{N\}\,\mathrm{Var}\,S_i. \end{aligned} \quad (4.1.5)$$

For S_i the jth moments are

$$E\{S_i^j\} = \exp\{j\mu + \tfrac{1}{2}j^2\sigma^2\}. \quad (4.1.6)$$

Exercise 4.1.2

Prove (4.1.6), then find $\mathrm{Var}\{S_i\}$.

Use of the Poisson model for the distribution of the number of pools and (4.1.5) gives

$$E\{Z\} = \lambda \exp\{\mu + \tfrac{1}{2}\sigma^2\}, \quad (4.1.7a)$$

$$\mathrm{Var}\{Z\} = \lambda \exp(2\mu + 2\sigma^2). \quad (4.1.7b)$$

Use of the negative binomial model for the distribution of the number of pools gives

$$E\{Z\} = \frac{\nu}{\alpha} \exp\{\mu + \tfrac{1}{2}\sigma^2\}, \quad (4.1.8a)$$

$$\mathrm{Var}\{Z\} = \frac{\nu}{\alpha^2} \exp\{2\mu + \sigma^2\} + \frac{\nu}{\alpha} \exp(2\mu + 2\sigma^2). \quad (4.1.8b)$$

As $\nu \to \infty$ the negative binomial model should collapse to a Poisson model. To see that it does, let $\nu, \alpha \to \infty$ in such a way that $\nu/\alpha \equiv \lambda$ is constant. Then $\nu/\alpha^2 = \lambda/\alpha \to 0$. Hence, equations (4.1.8) approach equations (4.1.7) as $\nu, \alpha \to \infty$.

4.1. EXPLORATION FOR EXHAUSTIBLE RESOURCES

Table 4.4.2

Parameter Estimates for Oil Exploration[a]

Parameter	Estimate	95% confidence region
v/α	0.1500	0.0382–0.2618
μ	8.967	7.7120–10.2222; 8.292–9.642
σ^2	4.2060	2.5439–8.8237

[a] From Uhler and Bradley (1970).

Uhler and Bradley also obtained maximum likelihood estimates of $M \equiv v/\alpha$, and σ^2. Their results are shown in Table 4.4.2. The first confidence region for μ corresponds to using $\sigma^2 = 8.8237$; the second to using $\sigma^2 = 2.5439$.

Knowing the distribution of resource pools in a certain region is not sufficient for optimizing the exploitation of the resource. One must know where the pools are! Hence, the problem of *search* becomes important. The rest of this section contains an introduction to elementary search theory (Koopman, 1980) as it can be applied to resource exploration. Cozzolino (1977, 1979), Cozzolino and Falconer (1977), and Menard and Sharman (1975) discuss some problems of search in petroleum exploration.

Consider first the case in which there is only one resource deposit in the region of area A. Let $q(a)$ be the probability that the deposit has not been found after an area a has been explored. Let $(\psi/A)\,da$ be the probability of discovering the deposit if area da is explored. Here ψ is a constant that is known a priori. The motivation of this assumption is that da/A is (approximately) the probability that the uniformly distributed resource deposit is in area da and ψ is the conditional probability of finding it, given that it is there. The law of total probability gives

$$q(a + da) = q(a)[1 - (\psi/A)\,da]. \tag{4.1.9}$$

Equation (4.1.9) leads to a differential equation for q:

$$\frac{dq}{da} = -\frac{\psi}{A} q. \tag{4.1.10}$$

Since $q(0) = 1$ (no exploration means no discovery),

$$q(a) = \exp\{-(\psi/A)a\}. \tag{4.1.11}$$

Suppose that area is explored at a constant rate. Then set $\psi a/A = \beta t$, where β is constant and t is time, i.e., time and area are linearly related. Equation (4.1.11) becomes, with $\hat{q}(t) = q(a)$,

$$\hat{q}(t) = \exp\{-\beta t\}. \tag{4.1.12}$$

Equation (4.1.12) is known as the "random search" formula (Koopman, 1980). It dates back to World War II and has proved to be quite useful in a variety of settings if one is willing to reinterpret β. Stone (1975) and Washburn (1981) discuss some uses and generalizations of these "exponential" search formula.

An alternative to the random search formula is the formula associated with exhaustive search (this idea also goes back to World War II). That is, assume that the region of interest is covered with perfectly navigated tracks so that there is no overlap in the search process. (Observe that there is still a chance of missing the deposit since ψ is the conditional probability of detecting the deposit given that it is present). If $\psi = 1$, then it is easily seen that for exhaustive search (4.1.12) is replaced by

$$\hat{q}(t) = \begin{cases} 1 - \beta t, & 0 \leq t \leq 1/\beta, \\ 0, & t > 1/\beta. \end{cases} \quad (4.1.12')$$

Exercise 4.1.3

Compare the two results (4.1.12) and (4.1.12'). Recall that the formula (4.1.12') was derived under the assumptions of (1) $\psi = 1$ and (2) perfect navigation. How must (4.1.12') be modified if $\psi < 1$? What would happen to (4.1.12') if navigational errors were present? [The question of navigational errors is actually quite difficult; see Stone (1975)].

When there are N resource deposits in the region, one can assume that (4.1.12) describes the probability of not discovering a single resource deposit and that search can be modeled as sampling without replacement. Then the probability of finding k deposits is given by the binomial formula

$$\Pr\{\text{finding } k \text{ resource deposits in } (0, t)\}$$

$$= \begin{cases} \binom{N}{k} \{1 - e^{-\beta t}\}^k \{e^{-\beta t}\}^{N-k}, & 0 \leq k \leq N, \\ 0, & \text{otherwise.} \end{cases} \quad (4.1.13)$$

If N is large and βt is small, so that $1 - e^{-\beta t} \ll 1$, the Poisson approximation to (4.1.13) gives

$$\Pr\{\text{finding } k \text{ deposits in } (0, t)\}$$

$$\simeq \frac{\exp\{-N(1 - e^{-\beta t})\}[N(1 - e^{-\beta t})]^k}{k!}. \quad (4.1.14)$$

A Taylor expansion of $1 - e^{-\beta t}$ shows that (4.1.14) can be approximated by a Poisson process with parameter $N\beta t$. The difficult part in most applied

4.1. EXPLORATION FOR EXHAUSTIBLE RESOURCES

problems is that N itself is not known. Ways of estimating N are discussed in Section 5.2.

Exercise 4.1.4

Critique the assumptions and approximations leading to this formula.

Now suppose that there are many regions to explore, say, M of them, and that each region contains either one deposit or no deposits (Fig. 4.1.1). Any large region can be put into this framework by dividing it into subregions that are sufficiently small. Each subregion can be called a "cell."

Let p_j be the prior probability that subregion j contains a resource pool. Assume that p_j is known from surveys or some other exogenous source of information. Using (4.1.12) show that the expression

$$p_j(1 - e^{-\beta_j t_j}) \tag{4.1.15}$$

is the probability of discovering resource in subregion j if time t_j is spent in exploration of that subregion.

A typical operational problem is the following one. Suppose that T hours are available for exploration. What is the optimal way to allocate effort in exploring the cells of the region? That is, how much effort t_j should be

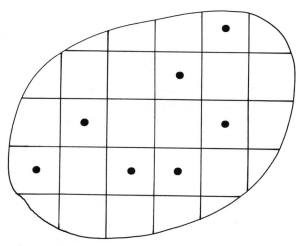

Fig. 4.1.1 An Illustration of the multicell exploration problem. The entire region is divided into a number of small cells. Each cell contains at most one deposit of the resource (denoted by ●). The problem is to allocate optimally the exploratory effort over the entire region.

allocated to cell j such that $\sum_{j=1}^{M} t_j = T$ and the number of deposits discovered is maximized? This question can be formulated as a mathematical programming problem of the following sort:

$$\text{maximize} \quad \sum_{j=1}^{M} p_j[1 - \exp(-\beta_j t_j)] \tag{4.1.16}$$

$$\text{such that} \quad \sum_{j=1}^{M} t_j = T, \quad t_j \geq 0.$$

The problem posed in (4.1.16) can be solved using either the Kuhn–Tucker theorem or the Pontryagin Maximum Principle (see Exercise 4.1.10). Another method of solution is to use the algorithm of Charnes and Cooper (1958). This algorithm provides considerable intuition into the nature of the solution; for this reason, it is worth discussing.

To see how this algorithm is motivated, first assume that T is small; then $1 - \exp(-\beta t_j) \approx \beta t_j$ and the functional in (4.1.16) is approximated by

$$\sum_{j=1}^{M} p_j[1 - \exp(-\beta_j t_j)] \simeq \sum_{j=1}^{M} p_j \beta_j t_j. \tag{4.1.17}$$

This suggests that for small T effort should initially go into the subregion where $p_j \beta_j$ is largest, since at this level of approximation the objective function is linear.

To help motivate the next step, set

$$P(t_j) = p_j[1 - \exp(-\beta_j t_j)], \tag{4.1.18}$$

so that $P(t_j)$ is the probability of finding a deposit with search effort t_j. Note that $dP/dt_j = p_j \beta_j \exp(-\beta_j t_j)$ is the marginal rate of increase of probability of discovery after an exploration time t_j in the subregion.

Based on (4.1.17) reorder the regions so that $p_1 \beta_1 > p_2 \beta_2 > p_3 \beta_3 > \cdots > p_m \beta_m$ (the case of some products being equal is left as an exercise). Equations (4.1.17) and (4.1.18) suggest that all search effort should go into subregion 1 until a time $T = t_{1s}$ such that the marginal rate of increase of probability in cells 1 and 2 is the same. Thus, t_{1s} is computed from the relationship

$$p_1 \beta_1 \exp(-\beta_1 t_{1s}) = p_2 \beta_2. \tag{4.1.19}$$

Solving (4.1.19) gives

$$t_{1s} = \frac{1}{\beta_1} \ln\left(\frac{p_1 \beta_1}{p_2 \beta_2}\right). \tag{4.1.20}$$

4.1. EXPLORATION FOR EXHAUSTIBLE RESOURCES

If $T > t_{1s}$, then effort should be split solely between subregions 1 and 2 until the rate of increase of probability of detection in subregion 2 matches the initial rate of increase in subregion 3. That is, effort is put solely into subregions 1 and 2 until effort t_{2s} is placed in subregion 2, where t_{2s} satisfies

$$p_2 \beta_2 \exp(-\beta_2 t_{2s}) = p_3 \beta_3, \tag{4.1.21}$$

so that

$$t_{2s} = \frac{1}{\beta_2} \ln\left(\frac{p_2 \beta_2}{p_3 \beta_3}\right). \tag{4.1.22}$$

How should this effort be split? All of the calculations done thus far suggest that effort above t_{1s} should be split between subregions 1 and 2 so that the *marginal* rate of increase of probability in each subregion is the same. Thus, if $T = t_{1s} + \delta$, the extra effort δ is split into $\delta = \delta_1 + \delta_2$, with effort δ_i going into subregion i in such a way that

$$p_2 \beta_2 \exp(-\beta_2 \delta_2) = p_1 \beta_1 \exp[-\beta(t_{1s} + \delta_1)]. \tag{4.1.23}$$

In light of (4.1.19), (4.1.23) becomes

$$\beta_2 \delta_2 = \beta_1 \delta_1. \tag{4.1.24}$$

Since $\delta = \delta_1 + \delta_2$ the values of δ_1 and δ_2 are

$$\delta_1 = \frac{\beta_2 \delta}{\beta_1 + \beta_2} = \frac{\delta}{\beta_1(1/\beta_1 + 1/\beta_2)},$$

$$\delta_2 = \frac{\beta_1 \delta}{\beta_1 + \beta_2} = \frac{\delta}{\beta_2(1/\beta_1 + 1/\beta_2)}. \tag{4.1.25}$$

Effort starts going into subregion 3 after effort t_{2s} is expended in subregion 2. Thus, if $\delta_2 = t_{2s}$ the value of δ must be

$$\delta = \left(\frac{\beta_2 + \beta_1}{\beta_1}\right) t_{2s} = \left(\frac{\beta_2 + \beta_1}{\beta_1 \beta_2}\right) \ln\left(\frac{p_2 \beta_2}{p_3 \beta_3}\right) \equiv \hat{t}_{2s}. \tag{4.1.26}$$

Thus, $\hat{t}_{2s}$ is the time above t_{1s} in which all effort is allocated to regions 1 and 2 only. Observe that $\hat{t}_{2s}$ can be written as

$$\hat{t}_{2s} = \left(\frac{1}{\beta_1} + \frac{1}{\beta_2}\right) \ln\left(\frac{p_2 \beta_2}{p_3 \beta_3}\right). \tag{4.1.27}$$

The decision rule derived thus far can be summarized as follows:

(1) If $T \le t_{1s}$, then all effort goes into subregion 1.
(2) If $t_{1s} < T \le t_{1s} + \hat{t}_{2s}$, then all effort goes into subregions 1 and 2. The effort $T - t_{1s}$ is split according to the rule (4.1.25).

The generalization of the results derived above proceeds as follows. All effort goes into subregions 1, 2, and 3 until effort t_{3s} is placed in subregion 3, where t_{3s} is given by

$$t_{3s} = \frac{1}{\beta_3} \ln\left(\frac{p_3 \beta_3}{p_4 \beta_4}\right). \tag{4.1.28}$$

Thus, if $T = t_{1s} + \hat{t}_{2s} + \delta$, the extra effort δ is split between subregions 1, 2, and 3 so that the marginal rates of increase of the probability of detection are the same. This gives

$$\beta_3 \delta_3 = \beta_2 \delta_2 = \beta_1 \delta_1, \tag{4.1.29}$$

where $\delta = \delta_1 + \delta_1 + \delta_2$ and δ_i is the effort placed into subregion i.

Exercise 4.1.5

Verify (4.1.29)

From (4.1.28) and (4.1.29), one concludes that effort t_{3s} can be placed in region 3 if a time $\hat{t}_{3s}$ above $t_{1s} + \hat{t}_{2s}$ is available, where

$$\hat{t}_{3s} = \left(\frac{1}{\beta_1} + \frac{1}{\beta_2} + \frac{1}{\beta_3}\right) \ln\left(\frac{p_3 \beta_3}{p_4 \beta_4}\right). \tag{4.1.30}$$

Exercise 4.1.6

Verify (4.1.30).

Thus, if $t_{1s} + \hat{t}_{2s} < T \le t_{1s} + \hat{t}_{2s} + \hat{t}_{3s}$, effort is placed into subregions 1, 2, and 3.

In general, define $\hat{t}_{ns}$ by

$$\hat{t}_{ns} = \left(\sum_{i=1}^{n} \frac{1}{\beta_i}\right) \ln\left(\frac{p_n \beta_n}{p_{n+1} \beta_{n+1}}\right) \tag{4.1.31}$$

and set $\hat{t}_{1s} = t_{1s}$. The general decision rule is then one of two cases. Recall that there are M subregions in total.

4.1. EXPLORATION FOR EXHAUSTIBLE RESOURCES

Case I:

First assume that

$$\sum_{n=1}^{M-1} \hat{t}_{ns} < T. \tag{4.1.32}$$

In this case some effort should be allocated to every region. An explicit formula can be derived as follows. It is clear that the effort allocated to subregion i will first reduce the marginal rate of increase of probability of detection in subregion i to that in subregion M, and then all rates will be reduced according to a generalization of (4.1.29). The effort needed for the first step, t_{iM}, is found by setting

$$p_i \beta_i \exp(-\beta_i t_{iM}) = p_M \beta_M \tag{4.1.33}$$

so that

$$t_{iM} = \frac{1}{\beta_i} \ln\left(\frac{p_i \beta_i}{p_M \beta_M}\right). \tag{4.1.34}$$

Exercise 4.1.7

How are the t_{iM} related to $\hat{t}_{ns}$ or $\hat{t}_{ns}$?

After effort t_{iM} is allocated to each subregion $i = 1, 2, \ldots, M - 1$, there remains effort Δ given by

$$\Delta = T - \sum_{i=1}^{M-1} t_{iM}. \tag{4.1.35}$$

This extra effort is split between the M cells according to

$$\beta_1 \Delta_1 = \beta_2 \Delta_2 = \beta_3 \Delta_3 = \cdots = \beta_M \Delta_M, \tag{4.1.36}$$

where $\Delta = \sum_{i=1}^{M} \Delta_i$ and Δ_i is the extra effort in subregion i. Thus, the solution of problem (4.1.16) in this case is

$$\begin{aligned} T_i^* &= t_{iM} + \Delta_i, \quad i = 1, \ldots, M-1, \\ t_M^* &= \Delta_M. \end{aligned} \tag{4.1.37}$$

Case II:

In this case, there is an M^* such that

$$\sum_{n=1}^{M^*-1} \hat{t}_{ns} < T \le \sum_{n=1}^{M^*} \hat{t}_{ns}. \tag{4.1.38}$$

Effort is allocated to the first M^* subregions only. The amount of effort allocated to subregion i is determined by equations analogous to (4.1.30)–(4.1.37). In particular, set

$$t_{iM^*} = \frac{1}{\beta_i} \ln\left(\frac{p_i \beta_i}{p_{M^*} \beta_{M^*}}\right), \tag{4.1.39}$$

$$\Delta = T - \sum_{i=1}^{M^*-1} t_{iM^*}. \tag{4.1.40}$$

Then the solution of problem (4.1.16) is given by

$$\begin{aligned} t_i^* &= t_{iM^*} + \Delta_i, \quad i = 1, 2, \ldots, M^* - 1, \\ t_{M^*} &= \Delta_{M^*}. \end{aligned} \tag{4.1.41}$$

Here $\Delta = \sum_{i=1}^{M^*} \Delta_i$, and the Δ_i satisfy

$$\beta_1 \Delta_1 = \beta_2 \Delta_2 = \cdots = \beta_{M^*} \Delta_{M^*}. \tag{4.1.42}$$

Exercise 4.1.8

Apply this theory to the following problem with five subregions:

Region	p_i	β_i (hr^{-1})
1	0.26	0.17
2	0.30	0.17
3	0.21	0.19
4	0.19	0.21
5	0.12	0.25

Assume that $T = 1, 4, 32$, and 256 hr, respectively, and solve the problem.

Exercise 4.1.9

Consider the following algorithm for solving problem (4.1.16).

(i) Ignore the nonnegativity constraint and solve the problem

$$\max_{t_j} \sum_{j=1}^{M} p_j[1 - \exp(-\beta_j t_j)] \quad \text{such that} \quad \sum_{j=1}^{M} t_j = T \tag{4.1.43}$$

using Lagrange multipliers.
(ii) If all the t_j are nonnegative, exit.
(iii) Delete those subregions in which the solution of (4.1.43) has $t_j < 0$. Return to step 1.

4.1. EXPLORATION FOR EXHAUSTIBLE RESOURCES

Show that this algorithm gives the same results as the one developed earlier in this section.

Exercise 4.1.10

(i) Solve the problem (4.1.16) by the Kuhn–Tucker theorem. How is the solution obtained this way related to the solution obtained by the Charnes–Cooper algorithm?

(ii) Solve the following problem using optimal control theory:

$$\text{minimize} \quad \int_a^b p(t) e^{-u(t)} \, dt$$

$$\text{such that} \quad u(t) \geq 0, \quad \int_a^b u(t) = U.$$

How is this problem related to the algorithm given in the text?

One interesting, and somewhat counterintuitive, result of this model is the following set of observations. The fundamental search model (4.1.11) is an exponential and thus is memoryless. Suppose that after effort t_{1s} is expended and no deposit is found, one then resolves the problem. Why wouldn't all the effort just go back into cell 1 until the deposit is found?

Exercise 4.1.11

Provide the answer to the question just posed before going on.

The answer to the question is that the "prior probability" changes as exploration occurs. In particular, observe that

$$p_1 = \Pr\{\text{deposit is in cell 1 before any exploration}\}. \quad (4.1.44)$$

Now assume that t_{1s} hours of exploration have been expended but that the deposit was not found. When one resolves problem (4.1.16), the prior probability p_1 is not used. Rather, the new prior probability is the posterior probability that a deposit is in cell 1 conditioned on no discovery after t_{1s} hours of search. This posterior probability is given by

$\Pr\{\text{deposit is in cell 1} \mid \text{no discovery after } t_{1s} \text{ hours of exploration}\}$
$= \Pr\{\text{deposit is in cell 1 and no discovery after } t_{1s} \text{ hours of exploration}\}/\Pr\{\text{no discovery}\}. \quad (4.1.45)$

The key here is the denominator in (4.1.45). In particular, it consists of two terms: a deposit may be present and not discovered [this event occurs with probability $p_1 \exp(-\beta t_{1s})$] or there may be no deposit in cell 1 (this event occurs with probability $1 - p_1$). Thus (4.1.45) becomes

Pr{deposit is in cell 1 | no discovery after t_{1s} hours of exploration}

$$= \frac{p_1 \exp(-\beta t_{1s})}{(1 - p_1) + p_1 \exp(-\beta t_{1s})}. \tag{4.1.46}$$

If p_1 were 1, so that the deposit were present with certainty, then the memoryless property of the exponential distribution would indeed apply. That it does not and that the probability associated with a deposit in cell 1 changes after exploration is seen from (4.1.46).

For the last model in this section, consider the case in which there may be many deposits in each subregion. Assume that the number of deposits N has a Poisson distribution with parameter λ. It appears reasonable, too, to assume that λ itself is unknown. To begin, one conditions on λ. Using (4.1.14) and conditioning on λ, the distribution of the discoveries in any cell is

Pr{finding k deposits with search effort $t | \lambda$}

$$\simeq \frac{[\lambda(1 - e^{-\beta t})]^k \exp\{-\lambda(1 - e^{-\beta t})\}}{k!}. \tag{4.1.47}$$

Thus

E{number of deposits found with search effort $t | \lambda$} = $\lambda(1 - e^{-\beta t})$. (4.1.48)

Now assume that λ has a gamma distribution with parameters v and α. Averaging (4.1.48) gives

E{number of deposits found with search effort t]

$$= (v/\alpha)(1 - e^{-\beta t}). \tag{4.1.49}$$

The static optimization problem of interest is now to maximize the number of deposits found in $[0, T]$. The mathematical problem is then

$$\text{maximize} \quad \sum_{j=1}^{M} \frac{v_i}{\alpha_j} [1 - \exp(-\beta_j t_j)]$$

$$\text{such that} \quad \sum_{j=1}^{M} t_j = T, \quad t_j \geq 0. \tag{4.1.50}$$

The problem posed in (4.1.50) is essentially the same as (4.1.16) and can be solved in the same manner. There could, however, exist an "updating point," $T_d < T$, at which the parameters v_i and α_i in each of the subregions explored up to T_d are updated based on the information obtained with search effort T_d.

Exercise 4.1.12

Derive updating formulas for the updated parameters v'_i and α'_i.

The updated parameters lead to a new optimization problem over the remaining time:

$$\text{maximize} \quad \sum_{j=1}^{M} \frac{v'_j}{\alpha'_j}(1 - \exp(-\beta_j t_j)]$$

$$\text{such that} \quad \sum_{j=1}^{M} t_j = T - T_d, \quad t_j > 0. \tag{4.1.51}$$

Exercise 4.1.13

Is there an optimal way to a priori choose T_d? (This is a good research problem.)

Bibliographic Notes

The algorithm for the optimization described in the text is from Charnes and Cooper (1958). The paper of Gilbert and Richels (1981) contains a nice example in which the economic value of search information is assessed. The book by Harbaugh, Doveton, and Davis (1977) contains many detailed examples showing how search theory can be successfully applied to oil exploration. The book by Newendorp (1975) treats petroleum exploration as a problem in decision analysis and shows how many of the ideas developed here can be applied. The chapter by Harris and Skinner (1982) in the book edited by Smith and Krutilla (1982) describes many applications of these ideas to mineral exploration. The search model introduced in the text is often called the "random search model" and is one of the simplest models. More complex models are discussed by Stone (1975) and Washburn (1981).

Mangel (1984) discusses some other kinds of search problems associated with mineral resources. Search models are also discussed by Griffiths (1966) and Gilbert (1976). Deterministic exploration models are used by Pindyck (1978) and Derzko and Sethi (1981). Some deterministic problems involving many resource deposits are discussed in Weitzman (1975) and Hartwick (1978). The general problems associated with analyzing spatially distributed resources can be called problems of "spatial statistics." Ripley (1977, 1981) provides a good introduction to spatial statistics.

4.2 UTILIZATION OF AN UNCERTAIN RESOURCE WITHOUT LEARNING

The next three sections are concerned with the optimal utilization of an uncertain resource under various assumptions about the nature of the uncertainty and learning. Throughout these sections consumption and extraction are considered to be the same process, so the terms "optimal consumption rate" and "optimal extraction rate" will be used interchangeably.

This section contains a version of a model from Loury (1978) for the utilization of a resource of unknown size. It is assumed that the distribution function for the initial stock size is completely known at the start of consumption. Consequently, there is no learning in this model. If the distribution itself changed as the stock was exploited, then there would be active learning in the model.

To begin development of the model, first consider the following deterministic problem. Let $c(t)$ denote the consumption rate at time t, ρ the discount rate, $u(c)$ the utility of, consumption, and $\tilde{X}(0)$ the initial value of the resource stock (known with certainty). The mathematical problem of interest is

$$\text{maximize} \quad \int_0^T e^{-\rho t} u(c(t))\, dt$$

$$\text{subject to} \quad \int_0^T c(t)\, dt = \tilde{X}(0). \tag{4.2.1}$$

A problem similar to this one was solved in Chapter 3. Recall that one adjoins a Lagrange multiplier λ by writing

$$\max_{c(t)} \int_0^T [e^{-\rho t} u(c(t)) - \lambda c(t)]\, dt + \lambda \tilde{X}(0) = 0. \tag{4.2.2}$$

The Hamiltonian is simply

$$\mathcal{H} = e^{-\rho t} u(c(t)) - \lambda c. \tag{4.2.3}$$

The first-order necessary condition $\partial \mathcal{H}/\partial c = 0$ gives

$$e^{-\rho t} u'(c(t)) = \lambda. \tag{4.2.4}$$

Unlike the problem in Chapter 3, the end time T in (4.2.2) may be finite. In general, the end time is unknown, so that the transversality condition $\mathcal{H}(T) = 0$ applies. For the Hamiltonian (4.2.3) the transversality condition becomes

$$e^{-\rho T} u(c(T)) - \lambda c(T) = 0. \tag{4.2.5}$$

Since $\mathcal{H}$ is independent of stock size, $\dot{\lambda} \equiv 0$, so that λ is constant. Equation (4.2.4) shows that the Hotelling condition still holds. It can be shown that $u''(c) < 0$ is sufficient to guarantee the existence of an optimal consumption profile.

4.2. AN UNCERTAIN RESOURCE WITHOUT LEARNING

Exercise 4.2.1

Justify the assertion that $u''(c) < 0$ is sufficient for the existence of an optimal control.

Exercise 4.2.2

Explicitly solve the problem posed in (4.2.2–4.2.5) for

(i) $u(q) = q^\gamma$, $0 < \gamma < 1$;
(ii) $u(q) = \ln(q + 1)$.

Now consider the case in which $\tilde{X}(0)$ is unknown, but assume that its distribution function is known. This means that one knows

$$\Pr\{\tilde{X}(0) \leq x_0\} = F(x_0). \tag{4.2.6}$$

Here $F(\cdot)$ is a given distribution function.

The amount of resource consumed up to a time t, $X(t)$, is computed by integrating the consumption rate. That is, $X(t)$ is given by

$$X(t) = \int_0^t c(s)\,ds. \tag{4.2.7}$$

The probability that any more of the resource can be consumed is the same as the probability that $\tilde{X}(0) > X(t)$; this probability is $1 - F(X(t))$. Hence, for the case of an unknown initial resource stock, the optimization problem over an infinite time horizon can be written as

$$\max_{c(t)} \int_0^\infty e^{-\rho t} u(c(t))[1 - F(X(t))]\,dt \tag{4.2.8}$$

subject to $dX/dt = c$; $X(0) = 0$.

The problem posed by (4.2.8) can be solved using optimal control theory. The Hamiltonian is

$$\mathcal{H} = e^{-\rho t} u(c(t))[1 - F(X(t))] + \lambda c(t). \tag{4.2.9}$$

The first-order condition and adjoint equation are

$$0 = \partial \mathcal{H}/\partial c = e^{-\rho t} u'(c(t))[1 - F(X(t))] + \lambda, \tag{4.2.10}$$

$$\dot{\lambda} = -\partial \mathcal{H}/\partial X = e^{-\rho t} u(c) F'(X(t)). \tag{4.2.11}$$

Solving (4.2.11), with $f(x) = F'(x)$, gives

$$\lambda(t) = \int_0^t e^{-\rho s} u(c(s)) f(X(s))\,ds. \tag{4.2.12}$$

Using (4.2.12) in (4.2.10) gives

$$e^{-\rho t}u'(c(t))[1 - F(X(t))] = -\int_0^t e^{-\rho s}u(c(s))f(X(s))\,ds. \quad (4.2.13)$$

Exercise 4.2.3

Set $c = \dot{X}$ in (4.2.13) and obtain an integrodifferential equation that determines $X(t)$. Let

$$u(c) = c^\gamma, \quad 0 < \gamma < 1,$$
$$F(X) = 1 - \exp(-X^2/2\sigma^2), \quad (4.2.14)$$

and write an algorithm to solve (4.2.13).

To go further, it is better to switch to a dynamic programming analysis. To do this, consider an infinite time horizon and define $J(x_0, s)$ by

$$J(x_0, s) = \max_c \int_s^\infty e^{-\rho s'}u(c(s'))[1 - F(X(s'))]\,ds'$$
$$\text{subject to}\quad \dot{X} = c, \quad X(s) = x_0. \quad (4.2.15)$$

Then, in the standard way, $J(x_0, s)$ is shown to satisfy

$$J(x_0, s) = \max_c\{e^{-\rho s}u(c(s))[1 - F(x_0)]\,ds + o(ds) + J(x_0 + c\,ds, s + ds)\}. \quad (4.2.16)$$

Equation (4.2.17) leads to the dynamic programming equation

$$0 = \max_c\{e^{-\rho s}u(c(s))[1 - F(x_0)] + cJ_{x_0} + J_s\}. \quad (4.2.17)$$

Setting $J(x_0, s) = e^{-\rho s}w(x_0)$ gives an ordinary differential equation for $w(x_0)$. This equation is

$$\rho w(x_0) = \max_c\{u(c)[1 - F(x_0)] + cw_{x_0}\}. \quad (4.2.18)$$

Equation (4.2.18) requires either an initial condition or a growth condition. The appropriate growth condition is

$$w(x_0) \to 0 \quad \text{as} \quad x_0 \to \infty. \quad (4.2.19)$$

This corresponds to the choice that when almost everything has been consumed, there is very little payoff from consumption.

From (4.2.19) the optimal consumption rule satisfies, at least formally,

$$c^* = u'^{-1}\{-w_{x_0}/[1 - F(x_0)]\}. \quad (4.2.20)$$

Here $u'^{-1}(\cdot)$ is the inverse function corresponding to $u'(c)$. Equations (4.2.18) and (4.2.20) lead to a nonlinear differential equation for $w(x_0)$. It can be explicitly solved, for example, if $u(c) = (1/\gamma)c^\gamma$.

Exercise 4.2.4

Derive the nonlinear equation for $w(x_0)$ by substituting (4.2.20) into (4.2.18). Then solve it for the particular case of $u(c) = (1/\gamma)c^\gamma$, $0 < \gamma < 1$. Find the optimal feedback consumption rule [see Loury (1978), p. 632].

There is no learning in this particular model, since knowledge of $F(x)$ is not changed at all as consumption occurs. The most extreme example of this occurs if $F(x) = 1 - e^{-\mu x}$. In this case

$$\Pr\{\text{remaining reserves exceed } x \,|\, \text{consumed an amount } s\}$$

$$= \frac{\Pr\{\tilde{X}(0) > x + s\}}{\Pr\{\tilde{X}(0) > s\}} = \frac{e^{-\mu(x+s)}}{e^{-\mu s}} = e^{-\mu x}. \qquad (4.2.21)$$

Exercise 4.2.5

If $F(x) = 1 - e^{-\mu x}$, show that the optimal consumption profile is constant.

Exercise 4.2.6

Reconsider the case in which $F(x)$ is arbitrary but given and assume that an optimal consumption policy c^* has been determined. Suppose that at some time t_1 an amount of resource X_1 has been consumed. Show that if one resolves the problem at time t_1 and obtains an optimal consumption policy c_1^*, valid for times greater than t_1, then $c_1^* = c^*$. That is, show that no revision of consumption ever occurs. Why would this be true?

Bibliographic Notes

Loury (1979) gives more details and interpretations of the model presented here. The traversality condition, when the end time is unknown (Eq. 2.7) is discussed by Clark (1976) and Kamien and Schwartz (1981). Discussions of the economics of learning, and the need for it, can be found in the papers by Arrow (1962) and Crawford (1973).

4.3 UTILIZATION OF AN UNCERTAIN RESOURCE WITH LEARNING

The model of the previous section can be faulted because there is no learning about the nature of the distribution (no updating). Learning is clearly an active part of the exploitation of an unknown resource. Hoel

(1978a,b) and Gilbert (1979) develop models in which learning is explicitly considered. This section contains versions of their mocels.

A simple heuristic that helps motivate these models is the following. Suppose that a deposit S can have only two sizes, S_1 and S_2, and that initially $\Pr\{\tilde{S} = S_1\} = \alpha$. Assume that $S_1 < S_2$ and that an amount of resource S_1 has been consumed. If there remains any more of the resource, then one knows with certainty that $\tilde{S} = S_2$; the distribution has changed from one in which $\Pr\{\tilde{S} = S_2\} = 1 - \alpha$ to one in which $\Pr\{\tilde{S} = S_2\} = 1$. On the other hand, if one consumes an amount S_1 and finds that nothing remains, then $\Pr\{S = S_1\} = 1$. The problem is to figure out the optimal way to reach the point at which S_2 has been consumed.

In the models of Hoel (1978a,b,c, 1979) there are two distinct resource deposits $\tilde{X}_1$ and $\tilde{X}_2$ that are exploited under the following assumptions.

(1) Deposit $\tilde{X}_1$ must be consumed before $\tilde{X}_2$. The size $\tilde{X}_1(0)$ and extraction costs of deposit 1 are known with certainty. Deposit $\tilde{X}_1$ is extracted first.

(2) The initial size $\tilde{X}_2(0)$ and extraction costs of deposit 2 are unknown. They are represented by a probability distribution. When deposit $\tilde{X}_1$ is extracted completely, knowledge of $\tilde{X}_2$ is gained with complete certainty.

(3) Extraction and consumption are viewed as the same activity.

A situation satisfying these assumptions could arise, for example, if the two deposits were layered and the properties of the second deposit could be determined imperfectly by sampling through the first deposit. Once the first deposit is exhausted, the properties of the second deposit can be determined perfectly. The main problem is then to determine the best time at which the first deposit should be exhausted and the associated consumption profile.

To mathematically frame this problem, define the following variables: $q(t)$, the extraction rate at time t; $u(q)$, the utility of consumption (extraction); $S(t)$, the total amount of resource remaining at time t; T, the deterministic time at which the first deposit is exhausted; a, the known cost of extracting the first deposit; and $\tilde{b}$, the unknown cost of extracting the second deposit. As an objective functional it is reasonable to take the total discounted utility. The optimization problem of interest is then

$$J = \max_{\substack{q(t) \\ T}} \left\{ \int_0^T e^{-rt}\{u(q(t)) - aq(t)\}\, dt \right.$$

$$\left. + \mathrm{E}\left\{ \int_T^\infty e^{-rt}\{u(q(t)) - \tilde{b}q(t)\}\, dt \right\} \right\} \qquad (4.3.1)$$

4.3. AN UNCERTAIN RESOURCE WITH LEARNING

subject to

$$\dot{S} = -q(t),$$
$$S(0) = \tilde{X}_1(0) + \tilde{X}_2(0), \qquad S(T) = \tilde{X}_2(0), \qquad (4.3.2)$$
$$S(t) \geq 0, \qquad q(t) \geq 0.$$

The expectation in (4.3.1) is taken over the distributions of $\tilde{X}_2(0)$ and $\tilde{b}$, which are the unknown size of the second deposit and the unknown cost of extracting the second deposit, respectively.

Exercise 4.3.1

Define $g(\tilde{X}_2(0), \tilde{b})$ by

$$g(\tilde{X}_2(0), \tilde{b}) = \max_{y(t)} \int_0^\infty e^{-rt}\{u(y(t)) - \tilde{b}y\}\, dt. \qquad (4.3.3)$$

Here $y(t)$ is the consumption rate of the second stock. Observe that by definition

$$\int_0^\infty y(t)\, dt = \tilde{X}_2(0). \qquad (4.3.4)$$

Let $y^*(t)$ be the optimal consumption rate. The exercise is to show that

$$\frac{\partial}{\partial \tilde{b}} g(\tilde{X}_2(0), \tilde{b}) < 0, \qquad \int_0^\infty \frac{\partial}{\partial \tilde{b}} y^*(t)\, dt = 0,$$
$$\frac{\partial^2}{\partial \tilde{b}^2} g(\tilde{X}_2(0), \tilde{b}) > 0. \qquad (4.3.5)$$

For the case in which the extraction technology is the same, $\tilde{b} = a$, and the only uncertainty is in the distribution of $\tilde{X}_2(0)$. In this case one can approximate $g(\tilde{X}_2(0), \tilde{b})$ by

$$E\{g(\tilde{X}_2(0), \tilde{b})\} \simeq g(E(\tilde{X}_2(0)), a)$$
$$+ \tfrac{1}{2} g_{xx}(E(\tilde{X}_2(0)), a)\, \mathrm{Var}\{\tilde{X}_2(0)\}, \qquad (4.3.6)$$

where $g_{xx}(\cdot)$ is the second derivative of $g(\cdot)$ with respect to $\tilde{X}_2(0)$.

Exercise 4.3.2

Verify (4.3.6) and show that $g_{xx} < 0$. What does this mean?

Now define $F(T)$ by

$$F(T) = \max_{q(t)} \int_0^T e^{-rt}\{u(q(t)) - aq(t)\}\, dt$$

such that $\int_0^T q(t)\, dt = \tilde{X}_1(0).$ \hfill (4.3.7)

Observe that $F(T)$ is the optimal value of utility associated with the extraction of the first deposit conditioned on an exhaustion date T.

By shifting integration variables one can rewrite (4.3.1) as

$$J = \max_T \left[F(T) + \max_q \mathrm{E}\left\{ e^{-rT} \int_0^\infty e^{-rs}\{u(q(s)) - \tilde{b}q(s)\}\, ds \right\} \right]. \quad (4.3.8)$$

Using (4.3.3), (4.3.8) can be rewritten as

$$J = \max_T [F(T) + e^{-rT}\, \mathrm{E}\{g(\tilde{X}_2(0), \tilde{b})\}]. \quad (4.3.9)$$

The problem is to find the optimal T, that is, the T that maximizes J. The optimal time satisfies (assuming an internal maximum and using elementary calculus)

$$F'(T) - re^{-rT}\, \mathrm{E}\{g(\tilde{X}_2(0), \tilde{b})\} = 0. \quad (4.3.10)$$

Equation (4.3.10) is the fundamental one for this analysis. There are many difficult calculations buried inside it; some of those difficulties will now be exposed. First note that (4.3.10) can be rewritten as

$$F'(T) = Ae^{-rT}, \quad (4.3.11)$$

where $A = \mathrm{E}\{g(\tilde{X}_2(0), \tilde{b})\}r$ is simply a constant. Figure 4.3.1 shows the two curves $A_1 e^{-rT}$ and $A_2 e^{-rT}$ for $A_1 > A_2$. Equation (4.3.11) shows that the optimal T is the point where $F'(T)$ intersects Ae^{-rT}.

Now consider $F'(T)$, by way of examples.

Fig. 3.1 Graphical solution of (3.11). This figure shows the right-hand side of (3.11), with $A_1 > A_2$.

4.3. AN UNCERTAIN RESOURCE WITH LEARNING

Exercise 4.3.3

The basic problem is to characterize T and $F'(T)$ in

$$F(T) = \max_q \int_0^T [u(q) - aq] e^{-rt} \, dt. \tag{4.3.12}$$

For simplicity, set $a \equiv 0$, $\tilde{X}_1(0) = S$, and $\dot{X} = q$. In this case

$$F(T) = \max \int_0^T u(\dot{X}) e^{-rt} \, dt$$

such that $\quad X(0) = 0, \quad X(T) = S.$ \hfill (4.3.13)

Consider two functional forms for $u(\dot{X})$. They are

$$u(\dot{X}) = \frac{1}{\gamma} \dot{X}^\gamma, \quad 0 < \gamma < 1, \tag{4.3.14}$$

and

$$u(\dot{X}) = \ln \dot{X}. \tag{4.3.15}$$

Solve the optimization problem posed in (4.3.13) using each of these forms for the utility function and therefore determine $F(T)$ explicitly.

In general, one can show that $F(T)$ has the properties that

$$F'(T) \to \infty \quad \text{as} \quad T \to 0, \tag{4.3.16}$$

$$F'(T) \to 0 \quad \text{as} \quad T \to \infty. \tag{4.3.17}$$

The graphical solution of (4.3.11) is shown in Fig. 4.3.2.

This result can now be used to consider the effects of learning about the distribution of $\tilde{X}_2(0)$ on the optimal time to exhaust $\tilde{X}_1$. Suppose, for

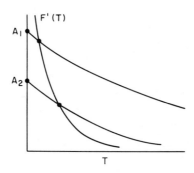

Fig. 3.2 Another graphical solution of (3.11). The solution of (3.11) is the value of T where $F'(T)$ intersects $A_i \exp(-rT)$.

example, that the distribution of $\tilde{X}_2(0)$ changes so that the mean remains the same but the variance increases. Since $g_{xx} < 0$ (Exercise 4.3.2), (4.3.6) shows that A in (4.3.11) decreases to first order, so that T^*, the solution of (4.3.11), increases. To see how T^* changes with A, and thus the distribution of $\tilde{X}_2(0)$, differentiate (4.3.11) with respect to A to obtain

$$F''(T^*)\frac{dT^*}{dA} = e^{-rT^*} - rAe^{-rT^*}\frac{dT^*}{dA}. \qquad (4.3.18)$$

Equation (4.3.18) shows that

$$\frac{dT^*}{dA} = e^{-rT^*}\{F''(T^*) + rAe^{-rT^*}\}^{-1}. \qquad (4.3.19)$$

The difficulty in applying (4.3.21) is that $F''(T)$ must be computed.

Exercise 4.3.4

Suppose that the distribution of the unknown extraction cost $\tilde{b}$ changes in such a way that the mean remains the same but the variance increases. What will happen to T^*, the solution of (4.3.11)?

Exercise 4.3.5

Observe that the problem posed in (4.3.8) is the same as the following optimization problem:

$$\max_{q(t), T}\left[\int_0^T e^{-rt}[u(q) - aq]\,dt + e^{-rT}\bar{g}\right], \qquad (4.3.20)$$

where

$$\bar{g} = E\{g(\tilde{X}_2(0), \tilde{b})\}. \qquad (4.3.21)$$

Show how to solve this problem using the Pontryagin Maximum Principle.

The model just discussed involves very little learning. The learning present in the model is passive in that once the first deposit is exhausted, complete knowledge of the distribution is available. Until that time, however, no control actions provide any information about the second deposit. A fruitful area for research is an extension of this model to include some exploration as well as extraction.

The model of Gilbert (1979) is more explicit in its treatment of learning. The simplest version of the model involves the case of a single resource stock which is costlessly extracted. The initial stock size $\tilde{X}(0)$ is unknown. To start, assume that $\tilde{X}(0)$ can only take a finite number of discrete values.

4.3. AN UNCERTAIN RESOURCE WITH LEARNING

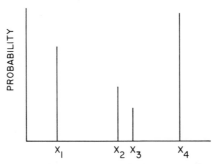

Fig. 3.3 Probability density for the case in which the deposit sizes are discretely distributed. Here X_i is the value associated with the ith deposit size.

The density function of $\tilde{X}(0)$ is then given by

$$f_0(x) = \sum_{j=1}^{N} \pi_j \delta(x - x_j). \qquad (4.3.22)$$

The distribution function of $\tilde{X}(0)$ is

$$F_0(x) = \sum_{j=1}^{N} \pi_j H(x - x_j). \qquad (4.3.23)$$

Here $\delta(s)$ is a delta function, $H(s)$ a step function, and $\sum_{j=1}^{N} \pi_j = 1$ with all $\pi_j > 0$.

According to (4.3.22) and (4.3.23), there are N possible choices for $\tilde{X}(0)$, that is, $\tilde{X}(0) = x_j$ with probability π_j for $j = 1, \ldots, N$. Figure 4.3.3 shows the density function.

The basic idea with this model is the following. As resource is consumed, one gains information about the stock level. Suppose that an amount x_1 has been consumed. If any more resource exists, then

$$\Pr\{\tilde{X}(0) = x_1 | \text{observations}\} = 0$$

and one must update the distribution of $\tilde{X}(0)$, assigning values to $\Pr\{\tilde{X}(0) = x_j | \text{observation}\}$ for $j = 2, 3, \ldots, N$. This explicit learning about the structure of the deposits can be included in the following way.

Define $\pi_j(i)$ by

$$\pi_j(i) = \Pr\{\tilde{X}(0) = x_j | \tilde{X}(0) > x_i\} \qquad (4.3.24)$$

so that $\pi_j(i)$ is the conditional distribution of $\tilde{X}(0)$ given that $\tilde{X}(0) > x_i$. Using Bayes theorem, $\pi_j(i)$ can be computed from (4.3.22) according to

$$\pi_j(i) = \pi_j \Big/ \sum_{j=i+1}^{N} \pi_j. \qquad (4.3.25)$$

In the denominator of this expression,

$$\sum_{j=i+1}^{N} \pi_j = \Pr\{\tilde{X}(0) > x_i\}.$$

As before, consumption and extraction are considered to be the same. Define

$$J(\tilde{X}(0)) = \max_{q(t)} E\left\{\int_0^\infty u(q(t))e^{-rt}\,dt\right\} \quad (4.3.26)$$

$$\text{subject to} \quad \int_0^\infty q(t)\,dt \leq \tilde{X}(0). \quad (4.3.27)$$

Here, as before, $q(t)$ is the extraction rate at time t and $u(q(t))$ the utility associated with an extraction rate $q(t)$. The problem is to find the optimal consumption rate, that is, the one that maximizes $J(\tilde{X}(0))$.

In order to solve this problem, proceed as follows. Suppose that an amount $X(\tau) = x_i$ has been extracted by time τ. An infinitesimal expenditure of effort (which is costless in this version of the model) will show either no resource (exhaustion) or at least an amount $\Delta_i \equiv x_{i+1} - x_i$.

Let T_i be the time (to be determined) required to consume optimally the amount Δ_i. Define

$$J_i = \max_q E\left\{\int_\tau^\infty u(q(t))e^{-rt}\,dt \,\Big|\, \tilde{X}(\tau) \geq \Delta_i\right\}. \quad (4.3.28)$$

The dynamic programming approach is now applied to (4.3.28) by breaking the range of integration $[\tau, \infty)$ into two pieces, $[\tau, \tau + T_i)$ and $[\tau + T_i, \infty)$. Unlike the treatment of previous DPEs, T_i will not be considered an infinitesimal. Instead, in this case T_i is a control variable as well; it is the amount of time used to consume optimally the amount of resource Δ_i. Following this prescription gives

$$J_i = \max_{[q(t)],\,T_i} E\left\{\int_\tau^{\tau+T_i} u(q(t))e^{-rt}\,dt + \int_{\tau+T_i}^\infty u(q(t))e^{-rt}\,dt \,\Big|\, \tilde{X}(\tau) \geq \Delta_i\right\}$$

$$(4.3.29)$$

Exercise 4.3.6

Justify, by using the principle of optimality, the argument leading to (4.3.29).

Define

$$J(\Delta_i, T_i) = \max_{q(t)} \int_\tau^{\tau+T_i} u(q(t))e^{-rt}\,dt$$

$$\text{such that} \quad \int_\tau^{\tau+T_i} q(t)\,dt = \Delta_i. \quad (4.3.30)$$

Note that $J(\Delta_i, T_i)$ can be computed deterministically.

Exercise 4.3.7

The problem posed in (4.3.30) is a free-terminal time problem. Solve it using optimal control theory with the state equation $\dot{X} = -q$, $X(0) = \Delta_i$. (The optimal end time T_i is determined from the transversality condition that the Hamiltonian vanishes at T_i).

Assume now that $J_i(\Delta_i, T_i)$ is known. Then (4.3.29) becomes

$$J_i = \max_{T_i} \left\{ J(\Delta_i, T_i) + \max_{q(t)} E\left\{ \int_{\tau+T_i}^{\infty} u(q(t))e^{-rt} dt \,\Big|\, \tilde{X}(\tau) \geq \Delta_i \right\} \right\}. \quad (4.3.31)$$

Consider the second expectation in more detail. It is given by

$$E\left\{ \int_{\tau+T_i}^{\infty} u(q(t))e^{-rt} dt \,\Big|\, \tilde{X}(\tau) \geq \Delta_i \right\} \quad (\text{set } t = y + T_i)$$

$$= E\left\{ \int_{\tau}^{\infty} u(q) \exp[-r(y + T_i)] dy \,\Big|\, \tilde{X}(\tau) \geq \Delta_i \right\}$$

$$= e^{-rT_i} E\left\{ \int_{\tau}^{\infty} e^{-ry} u(q) dy \,\Big|\, \tilde{X}(\tau) \geq \Delta_i \right\}. \quad (4.3.32)$$

Now either $\tilde{X}(\tau) = \Delta_i$, with probability π_i, or $\tilde{X}(\tau) > \Delta_i$, with probability $\alpha_{i+1} \equiv \sum_{j=i+1}^{N} \pi_j$. In the latter case, the expected gain is J_{i+1}, since if $X(\tau) > \Delta_i$, it must at least be Δ_{i+1}. In the former case, the expected gain is $\lim_{S \to 0} J(S)$, where $J(S)$ is given by (4.3.26) and (4.3.27).

Combining these results shows that J_i can be written as

$$J_i = \max_{T_i} \left\{ J(\Delta_i, T_i) + e^{-rT_i} \left[(1 - \alpha_{i+1}) \lim_{S \to 0} J(S) + \alpha_{i+1} J_{i+1} \right] \right\}. \quad (4.3.33)$$

Equation (4.3.33) is the fundamental DPE for this analysis. Although it is not easy to solve this equation, Gilbert (1979) shows that much can be learned about the extraction path and the properties of the value function. For most of the details, the reader is referred to Gilbert's paper. Sometimes, however, it is indeed possible to solve the DPE (4.3.33), as the following example shows.

Example 4.3.8

As a concrete example, consider the case in which $N = 2$ and

$$\begin{aligned} \Pr\{\tilde{X}(0) = X_1\} &= 1 - \pi, \\ \Pr\{\tilde{X}(0) = X_2\} &= \pi. \end{aligned} \quad (4.3.34)$$

For the utility function, choose $u(q) = q^{1-\varepsilon}$, with $0 < \varepsilon < 1$. It is then easily shown that

$$J(S) = S^{1-\varepsilon}(\varepsilon/r)^{\varepsilon},$$
$$J(S, T) = J(S)(1 - e^{-rT/\varepsilon})^{\varepsilon}. \quad (4.3.35)$$

In this case there is only one optimal time, T, which is the time at which the amount X_1 is consumed. The DPE (4.3.33) becomes

$$J_0 = \max_T \{J(X_1, T) + \pi e^{-rT} J(X_2 - X_1)\}. \quad (4.3.36)$$

Solving for T^*, the optimal value of T, and substituting back into (4.3.36) gives

$$J_0 = (\varepsilon/r)^{\varepsilon} \{X_1 + \pi^{1/(1-\varepsilon)}(X_2 - X_1)\}^{1-\varepsilon}. \quad (4.3.37)$$

Equation (4.3.37) provides a complete solution to the problem for two deposits.

Exercise 4.3.9

Find the size of a certain stock that would give the same utility as in (4.3.37). From this value compute the risk premium (see Section 3.1).

Exercise 4.3.10

Redo the previous example for $N = 3$ deposits and then do the general case of N deposits. What would the myopic Bayesian approach correspond to in this problem?

In reality, the deposits will not have a probability density function that consists of a set of spikes—there will be some variance associated with each deposit size. To see how this can be treated, replace the known deposits x_j with "smeared" deposits of mean x_j and variance σ_j^2 by replacing (4.3.22) by

$$f_0(x) = \sum_{j=1}^{N} \frac{\pi_j}{\sqrt{2\pi\sigma_j^2}} \exp\left\{\frac{-(x - x_j)^2}{2\sigma_j^2}\right\}. \quad (4.3.38)$$

Exercise 4.3.11

Show that $(4.3.38) \to (4.3.22)$ as all $\sigma_j \to 0$.

There are, of course, many interesting questions that could be asked of this new model. The entire preceding analysis, for instance, could be reproduced. Perhaps the most interesting question is the one that involves updating information. In the previous formulation, if the amount x_1 had been

4.4. OPTIMAL EXPLORATION AND EXPLOITATION

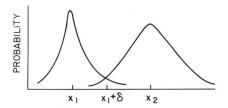

Fig. 3.4 Decision problem associated with a diffuse deposit. Shown are two deposit densities similar to (4.3.38). The problem is that when an amount $x_1 + \delta$ has been consumed, one must decide if the true density is the one on the left side or the one on the right side.

consumed and more resource was available, then the stock size was sure to be at least x_2. This need not be so in the new version. Suppose that an amount $x_1 + \delta$ has been consumed, as shown in Fig. 4.3.4.

In this situation a new decision problem is faced: given a consumption up to $x_1 + \delta$, how does one recalculate the distribution $f(x)$ and include it in the dynamic programming results to optimize consumption?

Exercise 4.3.12

If you can solve this problem, write a paper about your solution and publish it!

Bibliographic Notes

The papers of Hoel (1978a, b, c) contain many extensions of the simplest model presented in this section. Hoel is also concerned with the role of exogenous information. That is, suppose that one learns exogenously that the variance in $\tilde{X}_2(0)$ is decreased but the mean preserved. How does this affect the extraction profiles? The case of endogenous gains of information have not been treated. Hoel's monograph (Hoel, 1979) considers many additional problems, including the use of a backstop technology and details for linear and log-linear demand functions.

Gilbert's paper (Gilbert, 1979) treats extensions of the model presented in this section. These extensions include models for extraction costs, the certainty equivalent formulation, and the value of information calculations. The problem posed in Exercise 4.3.12 is a difficult one in statistical decision theory. The methods of sequential analysis (Wald, 1973), and pattern recognition (Duda and Hart, 1973) may turn out to be useful.

4.4 OPTIMAL EXPLORATION FOR AND EXPLOITATION OF AN EXHAUSTIBLE RESOURCE

Until now consumption, extraction, and exploration have been treated as the same activity. They clearly are not. The model developed in this section specifically separates consumption and exploration. This model was derived by Arrow and Chang (1980) and worked on by Chow (1981) and Derzko and Sethi (1981) as well.

When one considers exploration and exploitation separately, there are two kinds of variables. The first, the resource stock, is associated mainly with the exploitation phase. The second variable is a measure of exploration. In this context, imagine that the resource stock is located in a given region of known initial area. Exploration then corresponds to search over the region. Consequently, one can define the following variables. Let $Y_1(t)$ denote the known reserves of a resource at time t, $Y_2(t)$ the remaining area to be explored at time t, $R(t)$ the quantity of resource discovered up to time t, $q_1(t)$ the consumption rate at time t, and $q_2(t)$ the exploration rate at time t. Assume $Y_1(0)$, $Y_2(0)$, and $R(0)$ are given. Assume that exploration is a stochastic process characterized as follows:

$$\Pr\{\text{discovering a resource deposit in } (t, t+dt)|q_2(t)\}$$
$$= \lambda q_2(t)\, dt + o(dt),$$
$$\Pr\{\text{not discovering a resource deposit in } (t, t+dt)|q_2(t)\}$$
$$= 1 - \lambda q_2(t)\, dt + o(dt). \quad (4.4.1)$$

Exploration is thus a Poisson process with parameter $\lambda q_2(t)$. In most cases λ is probably not known, but initially it will be treated as if it were known.

With these assumptions, Y_1, Y_2, and R change according to the stochastic differential equations

$$dY_1 = -q_1(t)\,dt + d\pi(t), \qquad dY_2 = -q_2(t)\,dt, \qquad dR = d\pi(t). \quad (4.4.2)$$

In (4.4.2) $d\pi(t)$ is the jump process corresponding to exploration, that is,

$$d\pi(t) = \begin{cases} 1 & \text{with probability } \lambda q_2(t)\,dt + o(dt), \\ 0 & \text{with probability } 1 - \lambda q_2(t)\,dt + o(dt). \end{cases} \quad (4.4.3)$$

For the objective functional let $u(Y_1, Y_2, q_1, q_2, s)$ be the utility associated with the area and resource stock levels Y_2 and Y_1 and the consumption and exploration rates q_1 and q_2 at time s. Then set

$$J(y_1, y_2, t) = \max_{q_1, q_2} E\left\{\int_t^T u(Y_1(s), Y_2(s), q_1, q_2, s)\,ds \right.$$
$$\left. + \Phi(Y_1(T), Y_2(T), T)\,\bigg|\,Y_i(t) = y_i\right\}, \qquad i = 1, 2. \quad (4.4.4)$$

By standard methods it follows that $J(y_1, y_2, t)$ satisfies the DPE

$$\max_{q_1, q_2}\{u(y_1, y_2, q_1, q_2) + J_t - q_1 J_{y_1} - q_2 J_{y_2}$$
$$+ \lambda q_2[J(y_1 + 1, y_2, t) - J(y_1, y_2, t)]\} = 0. \quad (4.4.5)$$

4.4. OPTIMAL EXPLORATION AND EXPLOITATION

Exercise 4.4.1

Derive the DPE (4.4.5).

Performing the maximization in (4.4.5) gives (assuming interior maxima) conditions for the optimal exploration and consumption rates. These conditions are

$$\partial u/\partial q_1 = J_{y_1},$$
$$\partial u/\partial q_2 = J_{y_2} - \lambda[J(y_1 + 1, y_2, t) - J(y_1, y_2, t)]. \quad (4.4.6)$$

To go much further an explicit utility function is needed. Choose, for simplicity of calculation (also see Exercises 4.4.4, 4.4.6),

$$u(y_1, y_2, q_1, q_2) = \alpha_1 y_1 - \tfrac{1}{2}\beta_1 y_1^2 + \alpha_2 y_2 - \tfrac{1}{2}\beta_2 y_2^2$$
$$+ \omega_1 q_1 - \tfrac{1}{2}\eta_1 q_1^2 + \omega_2 q_2 - \tfrac{1}{2}\eta_2 q_2^2. \quad (4.4.7)$$

Exercise 4.4.2

Introduce matrices α, ω and vectors β, η and rewrite (4.4.7) in matrix notation.

Using (4.4.7), (4.4.6) becomes

$$\omega_1 - \eta_1 q_1 = J_{y_1}, \quad \omega_2 - \eta_2 q_2 = J_{y_2} - \lambda \, \Delta J(y), \quad (4.4.8)$$

where $\Delta J(y) \equiv J(y_1 + 1, y_2, t) - J(y_1, y_2, t)$. Thus, the optimal exploration and consumption rates are given by

$$q_1^* = \frac{1}{\eta_1}[-J_{y_1} + \omega_1],$$
$$q_2^* = \frac{1}{\eta_2}[\omega_2 + \lambda \, \Delta J(y) - J_{y_2}]. \quad (4.4.9)$$

Substituting these rates back into (4.4.5) gives the nonlinear partial differential equation

$$0 = \alpha_1 y_1 - \frac{1}{2}\beta_1 y_1^2 + \alpha_2 y_2 - \frac{1}{2}\beta_2 y_2^2 + \frac{\omega_1}{\eta_1}(\omega_1 - J_{y_1})$$
$$- \frac{1}{2\eta_1}(\omega_1 - J_{y_1})^2 + \frac{\omega_2}{\eta_2}(\omega_2 + \lambda \, \Delta J(y) - J_{y_2})$$
$$- \frac{1}{2\eta_2}(\omega_2 + \lambda \, \Delta J(y) - J_{y_2})^2 + J_t - \frac{1}{\eta_1}(\omega_1 - J_{y_1})J_{y_1}$$
$$- \frac{1}{\eta_2}(\omega_2 + \lambda \, \Delta J(y) - J_{y_2})J_{y_2} + \frac{\lambda \, \Delta J(y)}{\eta_2}(\omega_2 + \lambda \, \Delta J(y) - J_{y_2}). \quad (4.4.10)$$

Examination of (4.4.10) shows that there are terms $O(y_i^2)$, $O(y_iy_j)$, $O(y_i)$, and $O(1)$. This suggests seeking a solution of the form

$$J(y_1, y_2, t) = \tfrac{1}{2}[H_1(t)y_1^2 + H_2(t)y_2^2] + H_3(t)y_1y_2$$
$$+ h_1(t)y_1 + h_2(t)y_2 + m(t). \qquad (4.4.11)$$

If the solution takes the form (4.4.11), the partial derivatives in the y variables are

$$J_{y_1} = H_1y_1 + H_3y_2 + h_1,$$
$$J_{y_2} = H_2y_2 + H_3y_1 + h_2. \qquad (4.4.12)$$

With the ansatz (4.4.11), the difference operator in (4.4.10) takes the form [with the time dependence of $H_i(t)$ and $hi(t)$ suppressed]

$$\Delta J(y) = \tfrac{1}{2}[H_1(y_1 + 1)^2 + H_2y_2^2] + H_3(y_1 + 1)y_2 + h_1(y_1 + 1) + h_2y_2 + m(t)$$
$$- \{\tfrac{1}{2}[H_1y_1^2 + H_2y_2^2] + H_3y_1y_2 + h_1y_1 + h_2y_2 + m(t)\}. \qquad (4.4.13)$$

Equation (4.4.13) simplifies to give

$$\Delta J(y) = \tfrac{1}{2}H_1\{2y_1 + 1\} + H_3y_2 + h_1. \qquad (4.4.14)$$

Before substituting (4.4.12) and (4.4.14) into (4.4.10), collect terms and rewrite (4.4.10) as

$$0 = \alpha_1 y_1 - \tfrac{1}{2}\beta_1 y_1^2 + \alpha_2 y_2 - \tfrac{1}{2}\beta_2 y_2^2 + J_t$$
$$+ \frac{1}{2\eta_1}(\omega_1 - J_{y_1})^2 + \frac{1}{2\eta_2}(\omega_2 + \lambda \Delta J(y) - J_{y_2})^2 + J_t. \qquad (4.4.15)$$

Using (4.4.12) and (4.4.14) in this equation gives

$$0 = \alpha_1 y_1 - \tfrac{1}{2}\beta_1 y_1^2 + \alpha_2 y_2 - \tfrac{1}{2}\beta_2 y_2^2 + \tfrac{1}{2}[H'_1 y_1^2 + H'_2 y_2^2] + H'_3 y_1 y_2$$
$$+ h'_1 y_1 + h'_2 y_2 + m' + \frac{1}{2\eta_1}(\omega_1 - H_1 y_1 - H_3 y_2 - h_1)^2 \qquad (4.4.16)$$
$$+ \frac{1}{2\eta_2}(\omega_2 + \lambda H_1 y_1 + \tfrac{1}{2}\lambda H_1 + \lambda H_3 y_2 + \lambda h_1 - H_2 y_2 - H_3 y_1 - h_2)^2.$$

The unknown coefficients of the y_i in (4.4.11) can be found by setting coefficients of the powers of y_1, y_2 in (4.4.16) to zero. This gives

$$O(1): \quad 0 = m' + \frac{1}{2\eta_1}(\omega_1^2 + h_1^2 - 2\omega_1 h_1)$$
$$+ \frac{1}{2\eta_2}(\omega_2^2 + \tfrac{1}{4}\lambda H_1^2 + h_2^2 - 2\omega_2 h_2 - \lambda H_1 \omega_2 - \lambda H_1 h_2); \qquad (4.4.17a)$$

4.4. OPTIMAL EXPLORATION AND EXPLOITATION

$O(y_1)$: $\quad 0 = \alpha_1 + h_1' + \dfrac{1}{2\eta_1}(-2\omega_1 H_1 + 2h_1 H_1)$

$\qquad + \dfrac{1}{2\eta_2}(2\omega_2 \lambda H_1 + \lambda^2 H_1^2 + 2\lambda^2 H_1 h_1 - 2\lambda H_1 h_2);$ (4.4.17b)

$O(y_2)$: $\quad 0 = \alpha_2 + h_2' + \dfrac{1}{\eta_1}(-\omega_1 H_3 + H_3 h_1)$

$\qquad + \dfrac{1}{2\eta_2}(2\omega_2 \lambda H_3 + 2\lambda^2 H_3 h_1 - 2\lambda H_3 h_2 + \lambda H_3 H_1$

$\qquad - 2\omega_2 H_2 - \lambda H_1 H_2 - H_2 h_1 \lambda + 2H_2 h_2);$ (4.4.17c)

$O(y_1^2)$: $\quad 0 = -\dfrac{\beta_1}{2} + \dfrac{1}{2} H_1' + \dfrac{1}{2\eta_1} H_1^2 + \dfrac{1}{2\eta_2}(\lambda^2 H_1^2 + H_3^2);$ (4.4.17d)

$O(y_1 y_2)$: $\quad 0 = H_3' + \dfrac{1}{\eta_1} H_1 H_3 + \dfrac{1}{2\eta_2}(2\lambda^2 H_1 H_3 - \lambda H_1 H_2 - 2\lambda H_3^2);$ (4.4.17e)

$O(y_2^2)$: $\quad 0 = -\dfrac{\beta_2}{2} + \dfrac{1}{2} H_2' + \dfrac{1}{2\eta_1} H_3^2 + \dfrac{1}{2\eta_2}(\lambda^2 H_3^2 + 2H_2^2).$ (4.4.17f)

The six equations (4.4.17) are equations for the six unknowns H_1, H_2, H_3, h_1, h_2, and m. To obtain an end condition, assume that $\Phi(Y_1(T), Y_2(T), T)$ in (4.4.4) takes the form

$$\Phi(y_1, y_2, T) = \tfrac{1}{2}[\tilde{H}_1 y_1^2 + \tilde{H}_2 y_2^2] + \tilde{H}_3 y_1 y_2 + \tilde{h}_1 y_1 + \tilde{h}_2 y_2 + \tilde{m} \quad (4.4.18)$$

and then equate (4.4.18) and (4.4.11).

Exercise 4.4.3

Using the matrix notation derived in Exercise 4.4.1, convert the equations (4.4.17) to matrix equations and then solve them.

The ability to obtain as complete a solution as we did depends on the particular form for $u(y_1, y_2, q_1, q_2)$. For example, Arrow and Chang (1980) assume that $u(y_1, y_2, q_1, q_2) = \tilde{u}(q_1) - pq_2$ and are lead to a complicated free boundary value problem that was also studied by Derzko and Sethi (1981a, b).

Exercise 4.4.4

If $u(y_1, y_2, q_1, q_2) = \tilde{u}(q_1) - pq_2$, then the DPE (4.4.5) becomes

$$\max_{q_1, q_2}\{\tilde{u}(q_1) - pq_2 + J_t - q_1 J_{y_1} - q_2 J_{y_2} + \lambda q_2 \Delta J\} = 0.$$

Show that exploration is a "bang-bang" control. Then show that the (y_1, y_2) plane can be broken into portions in which exploration is zero and portions in which exploration is as large as possible. The boundary separating these regions is the unknown free boundary. What conditions on the value function will hold as one crosses this boundary?

Exercise 4.4.5

The parameter λ was treated as known in (4.4.4)–(4.4.18). In most cases of interest, λ itself will be unknown. Suppose that λ has a probability distribution associated with it. How will the above analysis change?

Exercise 4.4.6

How far can one proceed in the analysis if $u(y_1, y_2, q_1, q_2) = q_1^\alpha q_2^\beta$, $0 < \alpha < 1$ and $0 < \beta < 1$?

Another approach to the problem is to approximate the DPE (4.4.5). That is, the difference operator in (4.4.5) is the term that causes difficulty in the analysis. The basic idea is to try to find approximations for the difference operation.

In order to do this, first assume that each deposit has size δ instead of size 1. Then, instead of (4.4.5), the DPE is

$$\max_{q_1, q_2} \{u(y_1, y_2, q_1, q_2) + J_t - q_1 J_{y_1} - q_2 J_{y_2}$$
$$+ \lambda q_2 [J(y_1 + \delta, y_2, t) - J(y_1, y_2, t)]\} = 0. \quad (4.4.19)$$

Expanding (4.4.19) to first order in δ gives

$$\max_{q_1, q_2} \{u(y_1, y_2, q_1, q_2) + J_t - q_1 J_{y_1} - q_2 J_{y_2}$$
$$+ \lambda q_2 (\delta J_{y_1} + O(\delta^2))\} = 0. \quad (4.4.20)$$

Let $\lambda \to \infty$ and $\delta \to 0$ so that $\lambda \delta \to \mu$, a nonzero finite value. Then (4.4.20) becomes

$$\max_{q_1, q_2} \{u(y_1, y_2, q_1, q_2) + J_t - q_1 J_{y_1} - q_2 J_{y_2} + \mu q_2 J_{y_1}\} = 0. \quad (4.4.21)$$

Exercise 4.4.7

To what deterministic control problem does (4.4.21) correspond? When would (4.4.21) be a good approximation?

4.5. PRICE DYNAMICS AND MARKETS

Assuming that both $\lambda\delta$ and $\lambda\delta^2$ are kept in the expansion, (4.4.20) is replaced by

$$\max_{q_1,q_2}\{u(y_1, y_2, q_1, q_2) + J_t - q_1 J_{y_1} - q_2 J_{y_2} + \mu q_2 J_{y_1} + \tfrac{1}{2} q_2 D J_{y_1 y_1}\} = 0,$$

(4.4.22)

where $D = \lambda\delta^2$.

Exercise 4.4.8

To what stochastic control problem does (4.4.22) correspond? What can be said about its solution?

Bibliographic Notes

The free boundary problem that arises when a linear utility function is used was solved numerically by Derzko and Sethi (1981a). Derzko and Sethi (1981b) also consider the solution of the deterministic version of this problem [that is, (4.4.21)]. Free-boundary problems are very difficult mathematical problems. An introduction to such problems is found in Rubinstein (1971). The solution given in the text, which uses a quadratic utility function, follows the work of Chow (1981). It should be noted that in order to explicitly separate exploration and consumption, learning was suppressed. Learning in this model, as Chow (1981) points out, would involve assigning a probability distribution to λ and updating this distribution as exploration occurs. This problem is still unsolved. Duffie and Taskar (1983) and Hagan et al. (1981) study the diffusion and other approximations for the case of the linear utility function.

4.5 PRICE DYNAMICS AND MARKETS FOR EXHAUSTIBLE RESOURCES

In this section a problem of exhaustible resource management in the face of uncertain demand and fluctuations in reserves is considered. The model presented here is similar to the one used by Pindyck (1980). In this section consumption and extraction are treated as the same process.

Assume that the price per unit resource, $p(t)$, is given by

$$p(t) = y(t) f(q). \tag{4.5.1}$$

Here $f(q)$ is a demand function associated with a consumption rate $q(t)$, and $y(t)$ is a random process satisfying

$$dy/y = \alpha \, dt + \sigma_1 \, dW. \tag{4.5.2}$$

In (4.5.2) α is the "deterministic rate of interest," in the sense that $p(t)$ grows as $e^{\alpha t}$ in the absence of fluctuations. In (4.5.2) $W(t)$ is standard Brownian motion, with $E\{dW\} = 0$ and $E\{(dW)^2\} = dt$, and σ_1 is a known constant.

In this model reserves of the resource are accumulated in a stochastic fashion and exhausted at rate $q(t)$. Let $R(t)$ denote the reserves at time t. It is then assumed that

$$dR = -q\,dt + \sigma_2\,dW_2, \qquad R(0) = R_0, \qquad (4.5.3)$$

where R_0 is known exactly.

Exercise 4.5.1

According to (4.5.3), $R(t)$ could become negative. When would this assumption be valid?

The problem faced by the producer is then to maximize the net profit over $q(t)$:

$$\max_q E\left\{\int_0^T [yf(q) - C_1(R)]qe^{-\delta t}\,dt\right\}, \qquad (4.5.4)$$

where $C_1(R)$ is the cost of production when the stock level is R and δ the discount rate.

Define $J(y, r, t)$ by

$$J(y, r, t) = \max_q E\left\{\int_t^T [yf(q) - C_1(R)]qe^{-\delta t}\,dt \,|\, y(t) = y, R(t) = r\right\}, \quad (4.5.5)$$

It then follows that $J(y, r, t)$ satisfies the dynamic programming equation

$$\max_q \{[yf(q) - C_1(r)]qe^{-\delta t} + J_t - qJ_r + \alpha y J_y$$
$$+ \tfrac{1}{2}(\sigma_1^2 y^2 J_{yy} + \sigma_2^2 J_{rr})\} = 0. \quad (4.5.6)$$

To simplify notation we introduce the present value of profit defined by

$$\pi(t, r, q) = [yf(q) - C_1(r)]qe^{-\delta t}. \qquad (4.5.7)$$

Assuming an internal maximum in (4.5.6) gives that the optimal q, q^* satisfies

$$\partial \pi/\partial q = J_r. \qquad (4.5.8)$$

If one substitutes $\partial \pi/\partial q = J_r$ back into (4.5.6), the DPE becomes (as usual) a nonlinear partial differential equation. In general, it is not possible to solve the resulting equation. Instead, a different approach, analogous to the method of Arkin et al. (1966) discussed in Chapter 1 will be used. To employ this method, observe that if one thinks of the arguments in the

4.5. PRICE DYNAMICS AND MARKETS

value function as random variables, so that the value function is $J(Y, R, t)$, then the increment in the value function satisfies

$$\frac{1}{dt}\mathrm{E}\{dJ(Y, R, t)\} = \frac{1}{dt}\mathrm{E}\{J_t\, dt + J_y\, dY$$
$$+ \frac{1}{2} J_{yy}(dY)^2 + J_r\, dR + \frac{1}{2} J_{rr}(dR)^2\} + o(dt). \quad (4.5.9)$$

Using (4.5.9) the DPE (4.5.6) can be written as [ignoring terms that are $o(dt)$]

$$\pi(t, r, q) + \frac{1}{dt}\mathrm{E}\{dJ\} = 0. \quad (4.5.10)$$

Differentiating (4.5.10) with respect to r gives

$$\frac{\partial \pi}{\partial r} + \frac{1}{dt}\mathrm{E}\{d(J_r)\} = 0. \quad (4.5.11)$$

Now take the expectation in (4.5.8) and use it in (4.5.11) to obtain

$$\frac{\partial \pi}{\partial r} = -\frac{1}{dt}\mathrm{E}\left\{d\left(\frac{\partial \pi}{\partial q}\right)\right\}. \quad (4.5.12)$$

Equation (4.5.12) is a key result in this analysis. To interpret it, integrate both sides to obtain

$$\int_t^\infty \frac{\partial \pi}{\partial r}\, dt' = \mathrm{E}\left\{\frac{\partial \pi}{\partial q}\right\}. \quad (4.5.13)$$

The left-hand side of (4.5.13) is the value associated with holding a unit amount of resource between t and ∞; the right-hand side is the expected profit (to first order) from extracting it at the present time. On the optimal path, these two values are balanced. Equation (4.5.12) is thus an extention of the Hotelling condition.

Recall that Chapter 3 contained a characterization of the price dynamics of an exhaustible resource market. It will be done here also for the special case of $p = p(t)$ only, i.e., $f(q) \equiv 1$ in (4.5.1). Then

$$\pi = [p(t) - C_1(R)]q e^{-\delta t} \quad (4.5.14)$$

and

$$\partial \pi / \partial q = [p(t) - C_1(R)] e^{-\delta t}. \quad (4.5.15)$$

Since $p(t)$ and R are stochastic, so is $\partial \pi / \partial q$.

Using (4.5.15) in (4.5.12) gives

$$C_1'(r)q e^{-\delta t} = \frac{1}{dt}\mathrm{E}\{-\delta[p(t) - C_1(R)] e^{-\delta t}\, dt$$
$$+ e^{-\delta t}\, dp - e^{-\delta t}\, dC_1(R)\}. \quad (4.5.16)$$

In order to evaluate (4.5.16), note that
$$dC_1(R) = C_1'(R)\,dR + \tfrac{1}{2}C_1''(R)(dR)^2 + o(dt) \tag{4.5.17}$$
and that $E\{dR\} = -q\,dt$, $E\{(dR)^2\} = \sigma_2^2\,dt$.
Thus, the conditional expection of dC is
$$E\{dC_1(R)|R(t) = r\} = -qC_1'(r)\,dt + \tfrac{1}{2}C_1''(r)\sigma_2^2\,dt + o(dt). \tag{4.5.18}$$
Using (4.5.18) in (4.5.16), taking a second expectation, and rearranging gives
$$\frac{1}{dt}E\{dp\} = \delta\,E\{p(t) - C_1(R)\} + \tfrac{1}{2}\sigma_2^2 C_1''(r). \tag{4.5.19}$$

Equation (4.5.19) is the stochastic analog of the price dynamics determined in Chapter 3. In most cases [e.g., $C_1(R) = c_1/R$], $C_1''(r) > 0$, so that uncertainty tends to increase the rate of movement of the price of the resource.

Exercise 4.5.2

If $C_1''(r) = 0$, then (4.5.19) reduces to the deterministic result obtained in Chapter 3. Does this mean that all stochastic effects are unimportant? What can be said about the general role of uncertainty in this model?

Exercise 4.5.3

Do the same sort of calculations for the case in which price is a function of q as well, namely,
$$\pi = [p(q, t) - C_1(R)]qe^{-\delta t}. \tag{4.5.20}$$

Exercise 4.5.4

Consider this same problem for a monopolist in which
$$\pi = [MR - C_1(R)]qe^{-rt} \tag{4.5.21}$$
and solve it.

Exercise 4.5.5

In many situations the producer can simultaneously explore and extract. Pindyck (1980) considers two models for exploration. In the first model, exploration only provides information. Equation (4.5.3) is replaced by
$$dR = -q\,dt + \sigma_2(I)\,dW_2, \tag{4.5.22}$$

4.5. PRICE DYNAMICS AND MARKETS

where the information state I changes according to

$$dI = h(w)\, dt, \tag{4.5.23}$$

w is the exploration rate at time t, and $h(w)$ is a known function. What are reasonable properties of $\sigma_2(I)$? Assume that the cost of exploration is $c_2(w)$. Reformulate the optimization problem (4.5.4) and obtain as complete a solution as possible.

In the second model, exploration adds to the reserves as well. That is, if Z is the cumulative discovery of stock, then (4.5.3) is replaced by

$$dR = -q\, dt + f(w, Z, W_3)\, dt + \sigma_2\, dW_2, \tag{4.5.24}$$

where $f(w, Z, W_3)$ is presumed to be a known function and $W_3(t)$ is another Brownian motion process. What are reasonable properties of $f(w, Z, W_3)$? Reformulate the optimization problem and obtain as complete a solution as possible.

Bibliographic Notes

The procedure illustrated in this section allows one to learn about the solution of the DPE without actually solving it, which is generally a formidable task. The expansion (4.5.9) is essentially a use of the Ito calculus, as discussed in Chapter I of this book, Karlin and Taylor (1981), or Schuss (1980). Equation (4.5.12) is an extension of the classical, deterministic optimization rules for exploitation of an exhaustible resource [e.g., Clark (1976), Dasgupta and Heal (1979)]. Pindyck's paper (1980) contains a number of extensions and modifications of simple model described in this section. Pindyck has used the approach described in this section to study the production process when price is exogenous and stochastic (Pindyck, 1981) and to study the behavior of the firm under uncertainty (Pindyck, 1982).

5

Exploration, Observation, and Assessment of Renewable Resource Stocks

In this chapter the stochastic techniques developed earlier in the book are applied to the problem of renewable resource stock assessment. Questions of stock assessment are often faced by managers. Consequently, this entire chapter is devoted to *descriptive* methods associated with resource stock assessment. The next chapter is devoted to prescriptive optimization questions.

One area not discussed here is the work of Dixon and Howitt (1979a,b), who apply Kalman filtering theory to problems of forest management. These papers are an excellent application of the linear filtering theory to forest management. The basic assumption is that the evolution of timber in a forest satisfies discrete-time linear dynamics with noise present and that sampling can be modeled by a linear measurement function perturbed by noise. With these assumptions Kalman filtering can be succesfully applied to address questions of timber volume assessment as well as management.

5.1 INTRODUCTION TO THE THEORY OF LINE TRANSECTS

A standard method for estimating renewable stocks without harvesting them is the method of line transects. The discussion given here concentrates on the *analytic* aspects of the line transects. There are many *numerical* and

5.1. THE THEORY OF LINE TRANSECTS

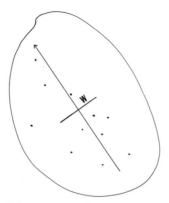

Fig. 5.1.1 The geometry of the transect problem. An observer walks a straight path and detects objects (denoted by dots) within a distance W of the path.

statistical questions that are not considered here. More details and references on the theory of transects can be found in the books by Burnham, Anderson, and Laake (1980) and Seber (1982), and an application to estimating whale populations can be found in the paper by Best and Butterworth (1980).

The setup is this. Imagine a large region containing many objects and that it is desired to estimate the density of the objects (Fig. 5.1.1). To estimate the density one follows a straight path in the region and counts the objects within a distance $2W$ of the path. Here W can be thought of as a detection width.

To provide a theoretical foundation for this operational problem proceed as follows. Let

$$g(x) = \Pr\{\text{detection at lateral (perpendicular)} \\ \text{range } x | \text{object is present}\}. \tag{5.1.1}$$

Exercise 5.1.1

What would the lateral range curve $g(x)$ be likely to look like, as a function of x?

The unconditional probability of detecting an object in the strip of width $2W$, P_W, is then found by integrating $g(x)$ over the strip. This gives

$$P_W = \frac{1}{2W} \int_{-W}^{W} g(x)\, dx = \frac{1}{W} \int_{0}^{W} g(x)\, dx. \tag{5.1.2}$$

The second equality in (5.1.2) follows from the assumption that $g(x)$ is an even function. The ultimate goal is to obtain an estimate for the density D

of objects per unit area. An estimator of D is found as follows. Suppose that many independent transects are carried out in the region of interest. If the total length of the transects is L, then the expected number of objects observed in the entire transect is

$$E\{n\} = D \cdot 2LW \cdot P_W. \tag{5.1.3}$$

Using (5.1.2) in (5.1.3) gives

$$E\{n\} = 2LD \int_0^W g(x)\,dx. \tag{5.1.4}$$

Let $f(x)$ be the conditional density for a detection at distance x, given that a detection occurred. This conditional density is given by

$$f(x)\,dx = \frac{\Pr\{\text{detection occurred in } (x, x+dx)\}}{\Pr\{\text{detection occurred at all}\}}$$

$$= [g(x)\,dx/W] \bigg/ \left[\frac{1}{W}\int_0^W g(x)\,dx\right], \tag{5.1.5}$$

so that $f(x) = g(x)/\int_0^W g(x)\,dx$.

Assuming that $g(0) = 1$, then using (5.1.5), (5.1.4) can be rewritten as

$$E\{n\} = 2DL/f(0). \tag{5.1.6}$$

The situation is now summarized as follows: D and $f(0)$ are unknown physical quantities and n is a random variable. One goes to the field and obtains a set of sighting distances, say, $X_1, X_2, \ldots, X_N$. D must be estimated from these observations. Since $f(0)$ is also unknown, the first problem of transect theory is to estimate $f(0)$ from a given set of sighting ranges $X_1, \ldots, X_N$. One could, in principle, try to estimate $f(0)$ in some nonparametric fashion. The more typical procedure is to choose a functional form for $f(x)$ that depends upon one or more parameters. Since $f(x)$ is a density, one must have that $\int_0^W f(x)\,dx = 1$. Some choices for $f(x)$ are shown in Table 5.1.1. Each choice depends upon at least one parameter. So, given a set of observations, the problem is now to estimate the parameters.

Exercise 5.1.2

Suggest another form for $f(x)$. For example, what about orthogonal polynomials with weighting factor e^{-x}?

Let $f(x, p)$ denote any of the entries in Table 5.1.1, where p denotes one or more parameters. Assume that $\mathbf{X} = (X_1, \ldots, X_N)$ is a set of observations and that the observations are independent identically distributed random

5.1. THE THEORY OF LINE TRANSECTS

Table 5.1.1

Suggested Choices for $f(x)$[a]

Form	Common name	Parameters to be estimated
$\dfrac{\lambda e^{-\lambda x}}{1 - e^{-\lambda W}}$	Negative exponential	λ
$\sum_{j=0}^{m} a_j x^i$	Polynomial	$a_0, a_1, \ldots, a_m$
$\dfrac{1}{W^*} + \sum_{k=1}^{m} \left\{ a_k \cos\left(\dfrac{k\pi x}{W^*}\right) + b_k \sin\left(\dfrac{k\pi x}{W^*}\right) \right\}$	Fourier series	$W^*, \{a_k\}, \{b_k\}$
$\dfrac{\exp\{-\frac{1}{2}(x/\sigma)^2\}}{\int_0^{W/\sigma} e^{-s^2/2} \, ds}$	Half normal	σ
$\exp\{-(ax + bx^2)\} / \int_0^W \exp\{-(ax + bx^2)\} \, dx$	Exponential polynomial	a, b
$\dfrac{\exp\{-(x/\lambda)^P\}}{\lambda \Gamma(1 + 1/P)}$	Exponential power series	λ, p
$\dfrac{ae^{-ax}}{(1 + be^{-ax}) \ln(1 + b)/b}$	Reversed logistic	a, b
$\dfrac{\beta}{\alpha}\left\{ 1 - \int_0^x \dfrac{\beta^\alpha y^{\alpha-1} e^{-\beta y}}{\Gamma(\alpha)} \, dy \right\}$	Incomplete gamma	α, β

[a] From Burnham, Anderson, and Laake (1980).

variables. The likelihood of such a set is the product of the density evaluated each X_i. This likelihood is given by

$$\mathscr{L}_N(\mathbf{X}; p) = \prod_{i=1}^{N} f(X_i, p). \tag{5.1.7}$$

One way to choose the parameters is to pick the value of p that maximizes the likelihood of the observations. The maximum likelihood estimate (MLE) for p satisfies (with ∇_p the gradient operator in the parameters) the equation

$$\nabla_p \mathscr{L}_N(\mathbf{X}; p) = 0. \tag{5.1.8}$$

If (5.1.8) is satisfied, then

$$\nabla_p \log \mathscr{L}_N(\mathbf{X}; p) = 0 \tag{5.1.9}$$

is also satisfied. It is often easier to work with (5.1.9) since it implies that

$$\nabla_p \left(\sum_{i=1}^{N} \log(f(X_i, p)) \right) = 0. \quad (5.1.10)$$

The solution of (5.1.10), denoted by $\hat{p}_N$, is an estimate of the true p. In general, the solution of (5.1.10) must be obtained by numerical methods. Some methods for the solution of nonlinear equations are discussed in Chapter 8. In addition to the MLE $\hat{p}_N$, one would like to obtain an estimate for the variance of $\hat{p}_N$, that is, an estimate for the quantity $E\{(p - \hat{p}_N)^2\}$. In addition to the classical methods for estimating the variance, one can use the Fisher information number (DeGroot, 1975) in the following way. Set

$$I(p) = \int \{\nabla_p \log f(x, p)\}^2 f(x, p) \, dx, \quad (5.1.11)$$

where ∇_p is the gradient operator with respect to the parameters. In the Bayesian view of inference, p is approximately normally distributed with mean $\hat{p}_N$ and variance $1/NI(\hat{p}_N)$.

Exercise 5.1.3

Compute the MLE estimates, classical variance, and Fisher information for the negative exponential and halfnormal forms of $f(x)$. It may be helpful to first consider the case in which W is infinite, since some of the calculations simplify. Then consider the case in which W is large, but finite.

Assume now that a form for $f(x)$ is chosen and that a transect is made. Suppose that n objects are discovered in the course of the transect. These data lead to a conditional estimate of $f(0)$, which we call $\hat{f}(0|n)$. In light of (5.1.6) a reasonable choice for the estimate of $\hat{D}$ is then found by solving (5.1.6) for D. This gives the estimate

$$\hat{D} \equiv n\hat{f}(0|n)/2L. \quad (5.1.12)$$

Exercise 5.1.4

Under what conditions is $\hat{D}$ an unbiased estimator of D?

In order to calculate the variance of $\hat{D}$, observe from (5.1.6) that $D = E\{n\}f(0)/2L$ so that

$$\begin{aligned} \text{Var}\{\hat{D}\} &= E\{(\hat{D} - D)^2\} \\ &= E\{(n\hat{f}(0|n) - E\{n\}f(0))^2\}/4L^2. \end{aligned} \quad (5.1.13)$$

5.1. THE THEORY LINE TRANSECTS

After a little simplification, (5.1.13) can be rewritten as

$$4L^2 \operatorname{Var}\{\hat{D}\} = E\{n^2 \hat{f}(0|n)^2\} - 2f(0) E\{n\} E\{n\hat{f}(0|n)\} + E\{n\}^2 f(0)^2.$$

(5.1.14)

From (5.1.6), $f(0) = 2DL/E\{n\}$. Now assume that the estimate $\hat{f}(0|n)$ has an asymptotic expansion of the form

$$\hat{f}(0|n) \simeq f(0) + k/n.$$

(5.1.15)

Exercise 5.1.5

Justify the choice of the expansion in (5.1.15). In this equation k is a random variable independent of n. It is assumed that $E\{k\} = 0$. Why is this assumption made?

Using (5.1.15) it can be seen that

$$\begin{aligned} E\{n\hat{f}(0|n)\} &\simeq E\{n\} f(0), \\ E\{n^2 \hat{f}(0|n)^2\} &\simeq E\{n^2\} f(0)^2 + E\{k^2\}. \end{aligned}$$

(1.5.16)

Using (5.1.16) in (5.1.14) and simplifying gives

$$\operatorname{Var} \hat{D} = (1/4L^2) f(0)^2 [\operatorname{Var}\{n\} + E\{k^2\}/f(0)^2].$$

(5.1.17)

But since $E\{n\} = 2LD/f(0)$, (5.1.17) becomes

$$\operatorname{Var} \hat{D} = \frac{Df(0)}{2L E\{n\}} \left[\operatorname{Var}\{n\} + \frac{E\{k^2\}}{f(0)^2} \right].$$

(5.1.18)

Equation (5.1.18) allows one to predict improvements in the accuracy of the estimate for increases in the length of the transect. It also shows how the various components of Var $\hat{D}$ are put together.

Exercise 5.1.6

It is reasonable to assume that the cost of a transect increases with its length. Formulate optimization problems for running a transect survey to achieve an estimate with a given level of accuracy. When does one know that n is large enough?

Exercise 5.1.7

In order to derive the results in this section, it was assumed that the X_i are measured perfectly. In most cases this will not be true, since the object sighted (a bird or school of fish) will not sit still and let the searcher measure the distance accurately. Thus, suppose that the measurements are given by $X_i = X_i^t + \varepsilon_i$, where X_i^t is the true value of the distance and ε_i a noise term. How would the analysis presented in this section change?

Bibliographic Notes

In addition to the monograph by Burnham, Anderson, and Laake (1980) and the paper by Best and Butterworth (1980), a good introduction to the transect method is found in the book by Seber (1982). A classic on the general field of sampling theory is the book by Cochran (1977). The concept of a lateral range curve for the detection process was introduced in World War II by scientists working in the Antisubmarine Warfare Operations Research Group. A history of the group, as well as the idea and uses of the lateral range curve, is found in Koopman's book on search theory (Koopman, 1980). Good introductions to the problems and techniques of parameter estimation are found in the books by Sorenson (1980) and Zacks (1971). The book by Zacks also discusses the Fisher information function and its uses. The paper by Butterworth (1982) contains a discussion on the form of $f(0)$.

Exercise 5.1.7 is an example of the "errors in variables" problem. Such problems are common in econometrics; the book by Judge *et al.* (1980) contains a good discussion of techniques for dealing with the errors in variables problem. Ludwig and Walters (1981) consider the errors in variables problem associated with fisheries management.

5.2 SURVEYS OF FISH STOCKS

The theory of transects is applicable for populations that can be seen easily. When surveying fish populations, there are often the added complications that the fish cannot be seen and that the fish may be hard to find. In this section three different aspects of the problem of finding and estimating fish stocks are treated. These three aspects are estimating the effectiveness of towed net samplers, the theory of effort allocation in surveys of stocks, and the effects of patchiness on survey accuracy.

EFFECTIVENESS OF TOWED-NET SAMPLERS

Barkley (1964) did a study of the effectiveness of towed-net samplers as survey devices. His work is an excellent example of operational analysis, showing how simple mathematics can be combined with a model of the system to learn a considerable amount about the system of interest.

Assume that a net of radius R is towed through the water at velocity U along the x axis and that when an organism sees the net it flees at speed u and direction θ. The geometry of this problem is shown in Fig. 5.2.1. The

5.2. SURVEYS OF FISH STOCKS

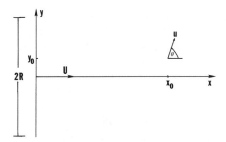

Fig. 5.2.1 The geometry of the towed-net sampler. The net travels with speed U; the animal flees at speed u. The width of the net is $2R$ and when the net is at the origin, the animal is at the point (x_0, y_0).

problem is to determine the effectiveness of such a towed-net sampler and to design the optimal towed-net sampler, e.g., the net that catches the most organisms at minimum relative cost. (The cost of towing a large net is greater than the cost of towing a smaller net due to drag forces on the net).

Exercise 5.2.1

What other design problems would be appropriate?

Suppose that the organism sees the net at time t_0, begins to flee, and that at time t' the net passes the organism; both are then a distance x' from the origin. Capture occurs if y', the y coordinate of the organism at t', is less than R (see Fig. 5.2.1).

The time t' at which the net overtakes the organism satisfies

$$U(t' - t_0) = x_0 + u(\cos \theta)(t' - t_0). \tag{5.2.1}$$

Here U is the speed of the net, u the speed of the organism, θ the direction in which the organism flees, and x_0 the initial lateral distance between the net and the organism. If y_0 is the initial value of the y coordinate for the organism, then at time t' the y coordinate of the organism is

$$y' = y_0 + u(t' - t_0) \sin \theta. \tag{5.2.2}$$

The condition for capture, $R > y'$, can be rewritten as

$$\frac{R}{x_0} > \frac{y'}{x_0} = \frac{y_0 + u \sin \theta (t' - t_0)}{(U - u \cos \theta)(t' - t_0)}. \tag{5.2.3}$$

If $u_{\max}$ is the maximum speed of the organism, then

$$R_{\min} = y_0 + \frac{x_0(u_{\max} \sin \theta)}{(U - u_{\max} \cos \theta)} \tag{5.2.4}$$

is the minimum net size needed to capture an organism starting at (x_0, y_0) and fleeing at angle θ. If it is assumed that θ is uniformly distributed on 0 to $\pi/2$, then the average of the minimum net size is

$$\langle R_{\min} \rangle = y_0 + \frac{2x_0}{\pi} \int_0^{\pi/2} \frac{u_{\max} \sin \theta}{U - u_{\max} \cos \theta} d\theta \qquad (5.2.5)$$

$$= y_0 + \frac{2x_0}{\pi} \ln\left(\frac{U}{U - u_{\max}}\right). \qquad (5.2.6)$$

An alternate approach is to find the escape speed u_e that ensures that an organism starting at (x_0, y_0) and fleeing at angle θ will escape a net of size R. Rearranging (5.2.3) gives for this speed

$$u_e = \frac{U}{[x_0/(R - y_0)] \sin \theta + \cos \theta}. \qquad (5.2.7)$$

Exercise 5.2.2

Use (5.2.7) to compute the minimum escape speed u_e^* as a function of the parameters U, R, x_0, and y_0. [*Hint*: Think of (5.2.7) as an equation parametrized by θ]. In order to design an effective towed-net sampler, one needs to characterize the dependence of u_e^* on R and U. What can be said about this dependence?

Exercise 5.2.3

Average (5.2.7) over θ and then sketch curves of $\langle u_e \rangle$ against R. To do this, let $r_0 = (x_0^2 + y_0^2)^{1/2}$ and parametrize the solution in terms of x_0 and r_0.

The next question to consider is how net speed U, ship speed S without a net, and net radius R are related. Barkley assumes that the net is towed at the maximum speed possible and that the drag caused by the net is proportional to $(UR)^2$. With these assumptions, one has

$$U = S - aR^2 U^2. \qquad (5.2.8)$$

It is now possible to approach the design question.

Exercise 5.2.4

Formulate and solve a problem for (i) the optimal net size given S; (ii) the optimal ship speed S given R. An assumption about the cost of operating a net of size R at speed S is needed. What assumption is reasonable?

ALLOCATION OF EFFORT IN SURVEYS OF STOCKS

The second problem considered is one that arises during the allocation of effort for surveys of fish stocks. Suppose that there are N separate regions that may contain fish and that a time T is available for surveying all of the regions. The basic operational problem is to find an allocation of times $\{t_1, \ldots, t_n\}$, with $t_i \geq 0$ being the time spent in region i and with $\sum t_i = T$, that is in some sense optimal. Different interpretations of "optimal" will give different allocations. One choice is to discover as many fish as accurately as possible. To see how this can be formulated mathematically, assume that if s hours are spent in region i, then a mean $M_i(s)$ and a variance $V_i(s)$ of the biomass of the population in the region will be obtained. A natural measurement of the accuracy of this biomass estimate is the coefficient of variation $CV_i(s) = \sqrt{V_i(s)}/M_i(s)$. If ω_i is a subjective measure of the value of an accurate estimate in region i, a nonlinear optimization problem that captures the essential features is

$$\text{minimize} \quad \sum_{i=1}^{N} \omega_i \, CV_i(t_i)$$

$$\text{subject to} \quad \sum_{i=1}^{N} t_i = T, \quad t_i \geq 0. \qquad (5.2.9)$$

Exercise 5.2.5

Justify the choice of differing ω_i. The optimization problem (5.2.9) is concerned with designing the most accurate survey possible. What other optimization criteria would lead to reasonable optimization problems?

In order to calculate CV_i, one needs a model relating survey effort and encounter rates. Two such models are the following.

(1) *Random encounters.* Assume that the encounter rate is a Poisson process with parameter λ. Thus $M_i = \lambda_i t_i$, $V_i = \lambda_i t_i$, and $CV_i = 1/\sqrt{\lambda_i t_i}$. Note that $t_i \to \infty$ and $CV_i \to 0$, so that, in principle, a survey of any desired accuracy could be conducted.

(2) *Patchy encounters.* Assume that the encounter rate is given by the negative binomial formula obtained by integrating the Poisson against a gamma distribution with parameters v_i and α_i, so that $M_i = (v_i/\alpha_i)t_i$, $V_i = M_i + (1/v_i)M_i^2$, and $CV_i = \sqrt{(\alpha_i + t_i)/v_i t_i}$. Note that, unlike the case of the Poisson model where $CV_i \to 0$ as $t_i \to \infty$, in this case $CV_i \to 1/\sqrt{v_i}$ as $t_i \to \infty$. The saturation represents a limit on accuracy due to patchiness. This result shows a decreasing marginal rate of improvement in the accuracy of

the survey. In particular, as soon as $t_i \gg \alpha_i$, there is little point in allocating more effort to the ith region, since the accuracy will not improve much with additional effort.

How are the parameters λ_i, v_i, and α_i to be estimated? If a region has never been surveyed, then the parameters must be guessed in the best way possible. If there were previous surveys in the region, then there is historical data that can be used to estimate the parameters in the following way. For the random encounter case, if N_0 schools were observed in the previous surveys of length T_0, the maximum likelihood estimate for λ is $\hat{\lambda} = N_0/T_0$.

In the patchy encounter case, the likelihood of observing N_0 schools in time T_0 given that the parameters are α and v is

$$\mathscr{L} = \frac{\Gamma(N_0 + v)}{\Gamma(v)} \frac{T_0^{N_0}}{N_0!} \alpha^v (\alpha + T_0)^{-(v+N_0)}. \tag{5.2.10}$$

Exercise 5.2.6

Compute the estimates of α and v in the following ways. First, use the method of moments, so that M_i and V_i are matched to the historical mean and variance. Next, compute the MLE estimates approximately as follows. One needs to solve the equations

$$\partial \mathscr{L}/\partial \alpha = 0, \tag{5.2.11}$$

$$\partial \mathscr{L}/\partial v = 0. \tag{5.2.12}$$

In order to do this, use the formula (valid for large x)

$$\Gamma(x) \simeq e^{-x} x^{x-1/2} \sqrt{2\pi} \tag{5.2.13}$$

to approximate the gamma function in the likelihood (5.2.10). When would it be easier, operationally, to use the method of moments rather than the MLE estimate?

The basic models are now set up. They can be used with the appropriate functional in (5.2.9) to determine the optimal level of effort in each subregion.

For the case of random encounters, the functional in (5.2.9) becomes

$$J = \frac{\omega_1}{\sqrt{\lambda_1 t_1}} + \frac{\omega_2}{\sqrt{\lambda_2 t_1}} + \cdots + \frac{\omega_N}{\sqrt{\lambda_N t_N}}, \tag{5.2.14}$$

$$= \frac{\omega_1}{\sqrt{\lambda_1}} \left\{ \frac{1}{\sqrt{t_1}} + \frac{\omega_2}{\omega_1} \sqrt{\frac{\lambda_1}{\lambda_2}} \frac{1}{\sqrt{t_2}} + \cdots + \frac{\omega_N}{\omega_1} \sqrt{\frac{\lambda_1}{\lambda_N}} \frac{1}{\sqrt{t_N}} \right\}. \tag{5.2.15}$$

5.2. SURVEYS OF FISH STOCKS

Defining $\beta_i \equiv \omega_i\sqrt{\lambda_1}/\omega_1\sqrt{\lambda_i}$, the problem is now to

$$\text{minimize} \quad 1/\sqrt{t_1} + \beta_2/\sqrt{t_2} + \cdots + \beta_N/\sqrt{t_N} \quad (5.2.16)$$

$$\text{subject to} \quad t_1 + t_2 + \cdots + t_N = T, \quad t_i \geq 0. \quad (5.2.17)$$

In order to solve this problem, the Kuhn–Tucker theorem (Avriel, 1976; Bronson, 1982) will be used. The form of the theorem used here is the following.

To solve the nonlinear program

$$\text{maximize} \quad J = F(\mathbf{X}), \quad \mathbf{X} \in R^n, \quad (5.2.18)$$

$$\text{subject to} \quad g_1(\mathbf{X}) \leq 0, \ldots, g_m(\mathbf{X}) \leq 0,$$

with $\mathbf{X} \geq 0$, introduce slack variables $x_{n+1}, \ldots, x_{2n+m}$ and form the Lagrangian

$$L \equiv F(X) - \sum_{i=1}^{n} \lambda_i(g_i(X) + x_{n+i}^2) - \sum_{i=m+1}^{m+2n} \lambda_i(-x_i + x_{n+i}^2). \quad (5.2.19)$$

Among the solutions to the Kuhn–Tucker conditions,

$$\partial L/\partial x_j = 0, \quad j = 1, 2, \ldots, 2n + m, \quad (5.2.20)$$

$$\partial L/\partial \lambda_i = 0, \quad i = 1, 2, \ldots, m + n, \quad (5.2.21)$$

$$\lambda_i \geq 0 \quad (5.2.22)$$

is the solution to the nonlinear program (5.2.18).

Exercise 5.2.7

Apply the Kuhn–Tucker theorem to (5.2.16, 17) and show that the Kuhn–Tucker conditions simplify to the following relation between the times:

$$1/t_1^{3/2} = \beta_2/t_2^{3/2} = \cdots = \beta_N/t_N^{3/2}. \quad (5.2.23)$$

In light of the first constraint (5.2.17), (5.2.23) implies that

$$t_1 = \frac{T}{1 + \beta_2^{2/3} + \cdots + \beta_N^{2/3}},$$

$$t_i = \beta_i^{2/3} t_1, \quad i = 2, \ldots, N. \quad (5.2.24)$$

Results of a calculation using (5.2.24) are shown in Table 5.2.1.

Table 5.2.1

Optimal Allocation of Survey Effort in the Case of Random Encounters[a]

Region	β_i	t_i
1	1	2.1
2	4	5.3
3	3	4.4
4	2.5	3.9
5	2.9	4.3
6	1.8	3.1
7	0.9	2.0

[a] $T = 25, J = 8.2$.

For the case of patchy encounters, the functional (5.2.9) becomes

$$J = \omega_1 \sqrt{\frac{\alpha_1 + t_1}{v_1 t_1}} + \omega_2 \sqrt{\frac{\alpha_2 + t_2}{v_2 t_2}} + \cdots + \omega_N \sqrt{\frac{\alpha_N + t_N}{v_N t_N}} \tag{5.2.25}$$

$$= \frac{\omega_1}{\sqrt{v_1}} \left\{ \sqrt{\frac{\alpha_1 + t_1}{t_1}} + \frac{\omega_2}{\omega_1}\sqrt{\frac{v_1}{v_2}}\sqrt{\frac{\alpha_2 + t_2}{t_2}} + \cdots + \frac{\omega_N}{\omega_1}\sqrt{\frac{v_1}{v_N}}\sqrt{\frac{\alpha_N + t_N}{t_N}} \right\}. \tag{5.2.26}$$

Defining $\beta_i = \omega_i \sqrt{v_1}/\omega_1 \sqrt{v_i}$, the nonlinear programming problem of interest is

$$\text{minimize} \quad \sqrt{\frac{\alpha_1 + t_1}{t_1}} + \beta_2 \sqrt{\frac{\alpha_2 + t_2}{t_2}} + \cdots + \beta_N \sqrt{\frac{\alpha_N + t_N}{t_N}}$$

$$\text{subject to} \quad t_1 + t_2 + \cdots + t_N = T, \quad t_i \geq 0. \tag{5.2.27}$$

Exercise 5.2.8

Apply the Kuhn–Tucker theorem and show that the optimal solution is characterized by

$$\frac{\alpha_1}{(\alpha_1 + t_1)^{1/2} t_1^{3/2}} = \frac{\beta_2 \alpha_2}{(\alpha_2 + t_2)^{1/2} t_2^{3/2}} = \cdots = \frac{\beta_N \alpha_N}{(\alpha_N + t_N)^{1/2} t_N^{3/2}}. \tag{5.2.28}$$

Show that this solution is unique (i.e., that each t_i is unique).

5.2. SURVEYS OF FISH STOCKS

One can easily imagine many simple, but important extensions of this model for survey design. For example, the cost of allocating effort to each region was not explicitly included in this model. A common problem is that often one is sampling many many species simultaneously, so that the objective functional is enlarged. An example of the multiple species and multiple area problem is found in the paper by Eggers (1979), which develops effort allocation schemes very similar to the ones developed in this section.

EFFECTS OF PATCHINESS ON SURVEY ACCURACY

The third problem of this section is a more explicit treatment of patchiness and its effect on surveys. This section contains a version of the model of Swierzbinski and Cain (1981). The simplest version of the problem arises in the following situation: S samples are taken in a region containing organisms at a background density λ_B and high concentration patches of organisms at density λ_p. Assume that each sample contains an area A_s, so that the number of organisms in the sample is A_s times the density of organisms in the sample (Fig. 5.2.2).

Now assume that when samples are taken, N_i organisms are observed in the ith sample. Assume that the samples are randomly distributed in the region. A simple estimator for the density of organisms is

$$\hat{\lambda} = \frac{1}{SA_s} \sum_{i=1}^{S} N_i. \qquad (5.2.29)$$

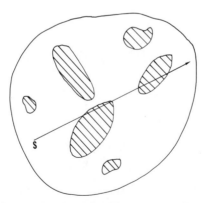

Fig. 5.2.2 The sampling problem in a patchy environment. The survey vessel (S) follows a path that takes it through the background density of organisms as well as through high-density patches (denoted by the hatched regions).

In the simplest case assume that the ith sample is completely in a patch or not. Let

$$f_i = \begin{cases} 1 & \text{if the } i\text{th sample is in a patch,} \\ 0 & \text{otherwise,} \end{cases} \quad (5.2.30)$$

and let $p = \Pr\{f_i = 1\}$.

The observed number of organisms in the ith sample is

$$N_i = f_i \lambda_p A_s + (1 - f_i)\lambda_B A_s = f_i(\lambda_p - \lambda_B)A_s + \lambda_B A_s. \quad (5.2.31)$$

Equation (5.2.31) can be used to compute the statistics of $\hat{\lambda}$. In particular, if the samples are independently taken, then the expected value of $\hat{\lambda}$ is

$$\mathrm{E}\{\hat{\lambda}\} = \frac{1}{SA_s} \sum_{i=1}^{S} \mathrm{E}\{N_i\} = \frac{S\,\mathrm{E}\{N_i\}}{SA_s} = p[\lambda_p - \lambda_B] + \lambda_B. \quad (5.2.32)$$

Exercise 5.2.9

Is $\hat{\lambda}$ an unbiased estimator (that is, is $\mathrm{E}\{\hat{\lambda}\} = 0$)?

For the variance of $\hat{\lambda}$, it follows from the binomial nature of f_i that

$$\mathrm{Var}\{\hat{\lambda}\} = \frac{1}{S^2 A_s^2} Sp(1 - p)(\lambda_p - \lambda_B)^2 A_s^2$$

$$= \frac{1}{S} p(1 - p)(\lambda_p - \lambda_B)^2. \quad (5.2.33)$$

Combining (5.2.33) and (5.2.32) gives the following for the coefficient of variation of $\hat{\lambda}$:

$$\mathrm{CV}\{\hat{\lambda}\} = \frac{\sqrt{[p(1 - p)/S](\lambda_p - \lambda_B)^2}}{p(\lambda_p - \lambda_B) + \lambda_B}$$

$$= \sqrt{\frac{p(1 - p)}{S}} \left(\frac{\lambda_p - \lambda_B}{p(\lambda_p - \lambda_B) + \lambda_B} \right). \quad (5.2.34)$$

Defining $\alpha = \lambda_B/\lambda_p$, the coefficient of variation becomes

$$\mathrm{CV}\{\hat{\lambda}\} = \sqrt{\frac{1 - p}{pS}} \frac{1 - \alpha}{1 - \alpha + \alpha/p}. \quad (5.2.35)$$

Equation (5.2.35) shows that the coefficient of variation of $\hat{\lambda}$ depends upon the fraction of patches p, the relative strength of the patches α, and the number of samples S. If $\alpha \simeq 1$, then the two densities are about the same and patchiness is not predominant. In this case $1 - \alpha$ is small. Typically, p will be small too,

5.2. SURVEYS OF FISH STOCKS

since it is a measure of the fraction of the total area in patches. If $O(1 - \alpha) \simeq O(p)$, then

$$\text{CV}(\hat{\lambda}) \simeq \sqrt{\frac{1-p}{pS} \frac{p^2}{p^2+1}} = \sqrt{\frac{p(1-p)}{S}} \frac{p}{1+p^2} \qquad (5.2.36)$$

and the coefficient of variation is small as soon as $S \simeq O(1)$.

The other extreme, a highly patchy environment, is the one in which α is small too, so that $O(\alpha) \simeq O(p)$. In such a case

$$\text{CV}(\hat{\lambda}) \sim \sqrt{\frac{(1-p)}{pS}} \frac{1}{2-p} \qquad (5.2.37)$$

and $\text{CV}(\hat{\lambda})$ will not be small until $S \simeq O(1/p)$. This indicates that a very large number of samples will be needed to achieve any precision [$S \simeq O(1/p)$ for CV's of 100%!]

Swierzbinski and Cain (1981) consider a variety of other questions, such as the effect of variability between patches and the determination of the optimal tow length.

Exercise 5.2.10

Consider three extensions of this simple model. First, repeat the analysis using the assumption that part of a tow may be in a patch and part of a tow may be out of any patch. Second, assume that patches are circles of radius r and analyze the case of linear tows through such circular patches. Third, assume that the cost of S tows of length L is

$$C(S, L) = (c_1 + c_2 L)S. \qquad (5.2.38)$$

Formulate and solve an optimal control problem concerning the number and length of tows.

Bibliographic Notes

General ideas about patchiness in ecological systems are ably discussed by Pielou (1977). Questions of patchiness and how to deal with it when allocating survey effort are some of the thorniest questions in ecology. Seber (1982) and Leaman (1981) address some of these issues. Introductions to the general analysis of spatial pattern are found in Bartlett (1964, 1974), Mollison (1977), and Ripley (1981). Descriptions of towed-net and other sampling methods are found in the books by Cushing (1981) and Everhart and Youngs (1981). The books by Cochran (1977) and Sukhatme and Sukhatme (1970) provide introductions to the problem of sampling many strata simultaneously. Exercise 5.2.10 is taken from Swierzbinski and Cain (1981). These kinds of models are also discussed by Swierzbinski (1984). Parameter estimates for the negative binomial distribution are discussed in Kendall and Stuart (1973).

5.3 MODEL IDENTIFICATION FOR AGGREGATING FISHERIES

In most resource problems one can develop a variety of models for the phenomena of interest. Often it happens that apparently similar models give quite dissimilar results. An example of this is provided by the models of the tuna purse-seine fishery developed by Clark and Mangel (1979). In such a situation, the question becomes not only one of fitting parameters of a model with the data, but also one of trying to use the data to select a model. If the predictions of the model have considerably different management implications, then it is important to continually try to identify which model appears to be the most appropriate. An approach to this problem is illustrated using the tuna models of Clark and Mangel (1979), which will now be described.

The tuna caught in the purse-seine fishery are captured at the surface of the ocean in association with flocks of birds, schools of porpoise, or sometimes debris. It is assumed that there is an underlying population that is not caught, and that the only information about the underlying population is obtained through information on the rate of captures at the surface of the ocean. One then needs to model the dynamics of the subsurface and surface populations and their interactions.

Imagine a given number K of identical surface "attractors." Tuna from the underlying population associate with these attractors according to some mechanism. Two models for the attraction are described below. It is assumed that the attractors are independent of one another and do not interchange associated tuna. Let N denote the number of tuna present in the backgound (subsurface) population. The number of tuna in an individual generic surface school is denoted by $Q = Q(t)$. In these models school occur only on the surface of the ocean.

In the first model tuna associate with a given attractor at a rate αN proportional to the background population and dissociate at a rate βQ proportional to the current school size. Thus the rate of change of the school population is

$$dQ/dt = \alpha N - \beta Q. \tag{5.3.1}$$

It is assumed that the dissociated tuna return to the background population. It will become apparent below that the key feature of (5.3.1) is that as the background population N decreases, the size of surface schools decreases (since if N is fixed, $dQ/dt < 0$ when $\alpha N < \beta Q$).

If N is held constant, the resulting steady-state school size Q^* is given by

$$Q^* = \alpha N/\beta. \tag{5.3.2}$$

5.3. MODEL IDENTIFICATION FOR AGGREGATING FISHERIES

If $Q(0) = Q_0$, (5.3.1) has the following solution for fixed N:

$$Q(t) = Q^*(1 - c_0 e^{-\beta t}), \qquad (5.3.3)$$

where $c_0 = 1 - Q_0/Q^*$. Thus, in this model at the steady-state the equilibrium size of schools is directly proportional to the background tuna population. (Since the number of attractors K is assumed fixed, the possibility that school size could also depend on K is not considered here). For ease of exposition, this will be called model A.

Exercise 5.3.1.

How do these results change if K is a function of time?

In the second model, assume that the maximum school size is a constant Q^* that is independent of the background tuna population. Equation (5.3.1) is replaced by the logistic-like equation

$$dQ/dt = \alpha N(1 - Q/Q^*), \qquad (5.3.4)$$

where Q^* is the maximum school size. The essential feature of this model is that $dQ/dt > 0$ if $Q < Q^*$, regardless of the value of N. Consequently, if $Q < Q^*$, surface schools grow even when the background population is very low.

Exercise 5.3.2

Contrast the preceding two models in predictions of the relationship between harvest rates and fishing effort.

If N is held constant, the solution of (5.3.4) is

$$Q(t) = Q^* - (Q^* - Q_0) \exp(-\alpha N t/Q^*). \qquad (5.3.5)$$

For ease of exposition, this will be called model B. Assume that the total catch rate is proportional to the school size on the surface. The catch rate Y is assumed to be given by

$$Y = bKEQ. \qquad (5.3.6)$$

Here b is a constant and E is fishing effort. Let $S = KQ$ be the total surface population. Then the dynamics for the total surface population are

$$\frac{dS}{dt} = \begin{cases} \alpha KN - \beta S - bES & \text{(Model A)}, \\ \alpha KN(1 - S/S^*) - bES & \text{(Model B)}, \end{cases} \qquad (5.3.7)$$

where $S^* = KQ^*$.

The first thing to do is to study the steady-state behavior of this model. Setting $dS/dt = 0$ gives the following catch equations for the steady-state catch rate Y (which is equal to bES):

$$Y = \begin{cases} b\alpha KEN/(\beta + bE) & \text{(Model A)}, \\ b\alpha KQ^*EN/(\alpha N + bQ^*E) & \text{(Model B)}. \end{cases} \quad (5.3.8)$$

For the two equations (5.3.8), observe that both models exhibit a saturation effect with respect to fishing effort E, but only model B exhibits a saturation effect with respect to tuna abundance N. For a fixed background population level N, the catch rate Y can be approximated for large and small E. For small E ignore the E in the denominator of (5.3.8) to obtain

$$Y \simeq \begin{cases} b\alpha NKE/\beta & \text{(Model A)}, \\ bQ^*KE & \text{(Model B)}. \end{cases} \quad (5.3.9)$$

Exercise 5.3.3

Show that the two equations in (5.3.9) are the same. (*Hint*: In model A, what is analogous to Q^*?)

For large E one obtains the following for both models:

$$\lim_{E \to \infty} Y = \alpha NK = Y_\infty. \quad (5.3.10)$$

For model B, in the case of fixed effort,

$$\lim_{N \to \infty} Y = bKQ^*E. \quad (5.3.11)$$

Next, one needs to adjoin a model for the subsurface population dynamics. As the model for the population dynamics of the subsurface tuna population, the logistic model can be used. Then the dynamics for $N(t)$ are assumed to be

$$dN/dt = rN(1 - N/\bar{N}) - \theta. \quad (5.3.12)$$

Here r is the intrinsic growth rate, $\bar{N}$ the environmental carrying capacity, and θ the net rate of transfer to the surface population.

The net rate of transfer θ is given by

$$\theta = \begin{cases} \alpha NK - \beta S & \text{(Model A)}, \\ \alpha NK(1 - S/S^*) & \text{(Model B)}. \end{cases} \quad (5.3.13)$$

5.3. MODEL IDENTIFICATION FOR AGGREGATING FISHERIES

The dynamic models of the entire fishery are then given by

Model A: $\quad dS/dt = \alpha KN - \beta S - bES,$

$\quad\quad\quad\quad\quad dN/dt = G(N) - (\alpha KN - \beta S);$ (5.3.14)

Model B: $\quad dS/dt = \alpha KN(1 - S/S^*) - bES,$

$\quad\quad\quad\quad\quad dN/dt = G(N) - \alpha KN(1 - S/S^*);$ (5.3.15)

where

$$G(N) = rN(1 - N/\bar{N}). \quad (5.3.16)$$

Exercise 5.3.4

There are six parameters $(\alpha, K, \beta, b, r, \bar{N})$ in (5.3.14) and five parameters (α, K, S^*, b, r) in (5.3.15). By introducing dimensionless variables, reduce the number of parameters in each set of equations as much as possible. In particular, show how the parameter $\varepsilon = rS^*/\alpha KN$ arises in a natural way upon scaling the equations. It is reasonable to assume that ε is small [Mangel (1982a) presents various parameter estimates leading to this conclusion]. If ε is small, show that there are two time scales in (5.3.14) and (5.3.15). Analyze the nondimensional equations and characterize the two time scales.

Although the difference between these two models may appear minor, their qualitative behavior turns out to be quite dissimilar. Their behavior is also quite different from the standard Schaefer model (Schaefer, 1957).

Figure 5.3.1 shows the solution trajectories $(N(t), S(t))$ for the system (5.3.14) for the two cases $\alpha K < r$ and $\alpha K > r$. The system has a unique, stable steady state at the point (N_∞, S_∞); the corresponding sustained yield from the fishery is given by $Y = bES_\infty$.

Exercise 5.3.5

Find the equations that (N_∞, S_∞) satisfy.

In Fig. 5.3.1 the effect of an increase in the effort parameter E is to rotate the isocline $\dot{S} = 0$ in a clockwise direction, thus decreasing both population levels, N_∞ and S_∞. The corresponding yield–effort curves are shown in Fig. 5.3.2.

The shape of the yield curves is explained as follows. Note from (5.3.13) that the constant $\rho = \alpha K$ represents the maximum net rate at which a unit of subsurface population aggregates to the surface; this may be referred to as the "intrinsic aggregation rate" (or "intrinsic schooling rate" in the present model). If the intrinsic aggregation rate ρ is less than the intrinsic

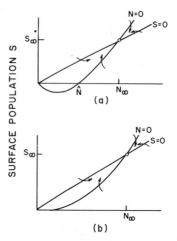

Fig. 5.3.1 Phase plane trajectories for Model A. A stable steady state exists at the point (N_∞, S_∞). In case (a) the intrinsic schooling rate is less than the intrinsic growth rate ($\alpha K < r$) and the population can never be depleted below N_∞. In case (b) the reverse is true ($\alpha K > r$) and the population can be driven to extinction.

growth rate r, then the population cannot be exhausted by the surface fishery; in this case $N \to N_\infty > 0$ and $Y \to Y_\infty > 0$ as effort $E \to \infty$. On the other hand, if $\rho > r$, then extinction is possible at sufficiently high levels of effort. This case is similar to the Schaeffer model. Equation (5.3.8) and its dynamical counterparts show that catch per unit effort (CPUE) is a seriously biased index of total stock abundance. The instantaneous CPUE is

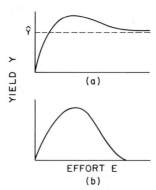

Fig. 5.3.2 Yield–effort curves for Model A. In case (a) the intrinsic schooling rate is less than the intrinsic growth rate and the yield decreases asymptotically as effort increases ($\alpha K < r$). In case (b) the intrinsic growth rate is less than the schooling rate and the yield goes to zero as effort increases to a level that drives the population to extinction ($\alpha K > r$).

5.3. MODEL IDENTIFICATION FOR AGGREGATING FISHERIES

simply an index of abundance for the surface population. Sustained CPUE overestimates the decline in abundance at high levels of effort. It is clear that in general no simple transformation of the CPUE index can provide an unbiased estimator of abundance for this model. Any fishery exploiting a substock of a biological population necessarily provides only partial information concerning total abundance; in the event that the fishery itself affects the relationship between the substocks, the interpretation of a time series of catch–effort data becomes extremely difficult. On the other hand, if the fishery affects the relationship between substocks, then management (i.e., control) actions can be used to provide information about the substocks and their relationships.

To summarize the results, if model A realistically represents the process of aggregation (via surface schooling) of tuna, then CPUE data may ultimately overestimate the decline in abundance of tuna. Management policy based on CPUE may then be unduly restrictive. The situation could be very different, however, if model B is the more realistic representation.

The solution trajectories of (5.3.15) are illustrated in Fig. 5.3.3, again corresponding to the cases $\alpha K < r$ and $\alpha K > r$.

If $\alpha K < r$ (case a), the system has a unique, stable steady state (N_∞, S_∞). As in model A, one has that $N \to N_\infty > 0$ as $E \to +\infty$. The yield–effort curve for this case has the same shape as for model A.

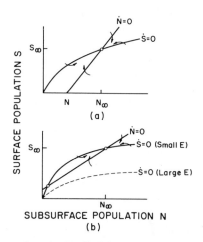

Fig. 5.3.3 Phase-plane trajectories for Model B. In case (a) the intrinsic schooling rate is less than the intrinsic growth rate and the model is very similar to Model A ($\alpha K < r$). In case (b) the intrinsic schooling rate is larger than the intrinsic growth rates ($\alpha K > r$). There are three steady states in this case: (N_∞, S_∞) and $(0, 0)$ are stable and there is an unstable steady state. As effort increases the unstable steady state and (N_∞, S_∞) approach each other and annihilate each other, leaving only $(0, 0)$ as the steady state.

A new phenomenon arises, however, in the case where $\alpha K > r$ (case b). For small E there now exist two stable steady states, at (N_∞, S_∞) and at $(0, 0)$, separated by an unstable steady state. As E increases the stable and unstable steady states coalesce and then disappear, leaving only the stable steady state at $(0, 0)$. In mathematical terminology, the system undergoes a "bifurcation" at the critical effort level $E = E_c$, where the two steady states coalesce. The graph of sustainable yield versus effort becomes multivalued for this case. Model B exhibits an explicit mathematical "catastrophe."

Exercise 5.3.4

The steady states can be computed by setting $dS/dt = dN/dt = 0$ in (5.3.15). Solve these equations and then verify analytically the results that were just derived by graphical means. Compute the value of E_c in terms of the biological parameters.

To confirm that such a "catastrophe" occurs in reasonable time for reasonable values of the parameters, Clark and Mangel performed a simulation study. Their results are shown in Figs. 5.3.4 5.3.5 (with $K = 5000$, $Q^* = 50$ tons, $b = 1 \times 10^{-4}$ per vessel day, $r = 1.5 \text{ yr}^{-1}$, and $\bar{N} = 10^6$ tons).

A fishery manager faced with the perplexing problem of managing such an aggregating fishery as this one must first decide if one model is more appropriate than the other. How can this be done? There are fundamentally two viewpoints. The first could be called a purely passive strategy. In this case data are accumulated, parameters are estimated, and a model is selected on the basis of the best fit of the model to the data. The second viewpoint is an active strategy in which a manager would try to affect controls that would also provide information about the system.

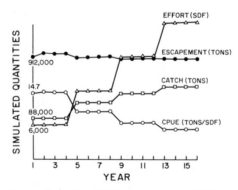

Fig. 5.3.4 Simulation results for Model B for the case in which the intrinsic schooling rate is less than the intrinsic growth rate.

5.3. MODEL IDENTIFICATION FOR AGGREGATING FISHERIES

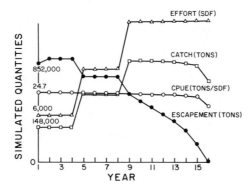

Fig. 5.3.5 Simulation results for Model B for the case in which the intrinsic schooling rate is greater than the intrinsic growth rate. Effort is increased in years 1, 5, and 9. The final level of effort produces a catastrophic decline in population. Note that this decline is not evident in catch per unit effort (CPUE) until the population is virtually eliminated.

To see how the passive strategy would proceed, consider the following intraseason situation. Assume that the data consist of (E_i, Y_i), where E_i is the effort and Y_i the yield in the ith week of the season. Assume that the steady-state approximation (5.3.8) to the catch rate is valid in each week. This equation can be rewritten as

$$Y = \begin{cases} EN/(p_1 + p_2 E) & \text{(Model A)}, \\ EN/(p_1 N + p_2 E) & \text{(Model B)}. \end{cases} \quad (5.3.17)$$

For Model A, $p_1 = \beta/b\alpha K$ and $p_2 = 1/\alpha K$; for Model B, $p_1 = 1/bKQ^*$ and $p_2 = 1/\alpha K$. Observe that one can only estimate the combinations of the fundamental parameters p_1 and p_2 from (5.3.17). In particular, one would need other means (e.g., surveys) to estimate the fundamental parameters separately. In each case there are three parameters to estimate; these are N, p_1, and p_2. If the season is short enough, then one can ignore growth during the season and set $N = N_0$, the value of N at the start of the season. After k weeks of fishing, the value of N is $N_0 - \sum_{i=1}^{k} Y_i$ and the projected steady-state catch in week $k + 1$ is

$$\hat{Y}_{k+1} = \begin{cases} \dfrac{E_{k+1}(N_0 - \sum_{i=1}^{k} Y_i)}{p_1 + p_2 E_{k+1}} & \text{(Model A)}, \\[2ex] \dfrac{E_{k+1}(N_0 - \sum_{i=1}^{k} Y_i)}{p_1(N_0 - \sum_{i=1}^{k} Y_i) + p_2 E_{k+1}} & \text{(Model B)}. \end{cases} \quad (5.3.18)$$

One way to estimate the parameter values (p_1, p_2, N_0) is by nonlinear least squares (Sorenson, 1980).

This suggests considering as least squares objective functionals for M weeks of data the following forms. For Model A,

$$J_A(p_1, p_2, N_0) = \sum_{k=1}^{M} \left[Y_k - \left(\frac{E_k(N_0 - \sum_{i=1}^{k-1} Y_i)}{p_1 + p_2 E_k} \right) \right]^2. \quad (5.3.19)$$

For Model B,

$$J_B(p_1, p_2, N_0) = \sum_{k=1}^{M} \left[Y_k - \left(\frac{E_k(N_0 - \sum_{i=1}^{k-1} Y_i)}{p_1(N_0 - \sum_{i=1}^{k-1} Y_i) + p_2 E_k} \right) \right]^2. \quad (5.3.20)$$

The least squares estimators for p_1, p_2, and N_0 in Model A satisfy the equations

$$\sum_{k=1}^{M} \left[Y_k - \left(\frac{E_k(N_0 - \sum_{i=1}^{k-1} Y_i)}{p_1 + p_2 E_k} \right) \right] \frac{E_k}{p_1 + p_2 E_k} = 0,$$

$$\sum_{k=1}^{M} \left[Y_k - \left(\frac{E_k(N_0 - \sum_{i=1}^{k-1} Y_i)}{p_1 + p_2 E_k} \right) \right] \frac{E_k(N_0 - \sum_{i=1}^{k-1} Y_i)}{(p_1 + p_2 E_k)^2} = 0, \quad (5.3.21)$$

$$\sum_{k=1}^{M} \left[Y_k - \left(\frac{E_k(N_0 - \sum_{i=1}^{k-1} Y_i)}{p_1 + p_2 E_k} \right) \right] \frac{E_k^2(N_0 - \sum_{i=1}^{k-1} Y_i)}{(p_1 + p_2 E_k)^2} = 0.$$

For Model B the estimates for p_1, p_2, and N_0 satisfy the equations

$$\sum_{k=1}^{M} \left[Y_k - \left(\frac{E_k(N_0 - \sum_{i=1}^{k-1} Y_i)}{p_1(N_0 - \sum_{i=1}^{k-1} Y_i) + p_2 E_k} \right) \right] \left[\frac{E_k}{p_1(N_0 - \sum_{i=1}^{k-1} Y_i) + p_2 E_k} \right.$$

$$\left. - \frac{E_k p_1 (N_0 - \sum_{i=1}^{k-1} Y_i)}{(p_1(N_0 - \sum_{i=1}^{k-1} Y_i) + p_2 E_k)^2} \right] = 0,$$

$$\sum_{k=1}^{M} \left[Y_k - \left(\frac{E_k(N_0 - \sum_{i=1}^{k-1} Y_i)}{p_1(N_0 - \sum_{i=1}^{k-1} Y_i) + p_2 E_k} \right) \right] \frac{E_k(N_0 - \sum_{i=1}^{k-1} Y_i)^2}{(p_1(N_0 - \sum_{i=1}^{k-1} Y_i) + p_2 E_k)^2} = 0,$$

$$\sum_{k=2}^{M} \left[Y_k - \left(\frac{E_k(N_0 - \sum_{i=1}^{k-1} Y_i)}{p_1(N_0 - \sum_{i=1}^{k-1} Y_i) + p_2 E_k} \right) \right] \frac{E_k^2(N_0 - \sum_{i=1}^{k-1} Y_i)}{(p_1(N_0 - \sum_{i=1}^{k-1} Y_i) + p_2 E_k)^2} = 0.$$

$$(5.3.22)$$

Once the parameter estimates are known, the residuals from the least squares can be used to help select a model. Unfortunately, it does not appear that the distribution of the residuals is easy to calculate; hence hypothesis testing cannot be done easily.

Exercise 5.3.7

Propose an alternative set of data that could be collected as well as an alternative procedure for selecting the models (e.g., what should be done if the data are yearly, rather than weekly?).

Exercise 5.3.8

Instead of the classical, nonlinear least squares approach, one could consider a Bayesian approach in which prior densities are given for p_1, p_2, and N_0. How would the model selection procedure work in this case?

There remains a whole set of questions such as are other models better, how does one assign confidence levels to the choice of model, and how does one manage to gain not only fish but information about the choice of model? These are fruitful areas for research.

The procedure for model selection described thus far is fundamentally passive. Observations are taken and used, but the system is not designed to maximize the information gained from the observations. Walters and Hilborn (1976) and Walters (1981) advocate an active strategy in which part of the management plan is to sample for information about the models by experimenting. In this case the harvest rate could be used as an experimental variable.

Example 5.3.9

Walters (1981) considers a model of the form

$$X(n + 1) = X(n) \exp[\alpha(1 - X(n)/B)]. \quad (5.3.23)$$

Here $X(n)$ is the population in year n and α and B are parameters. Assume that the two choices for models are $B = B_1$ or $B = B_2 > B_1$. The typical control variable, h, removes a fraction of the population, $0 < h < 1$, so that $X(n)$ in (5.3.23) is replaced by $(1 - h)X(n)$. If h is large so that $X(n)$ is small, then it should be clear that it may be essentially impossible to select one model over the other.

Exercise 5.3.10

Explain how the models in Example 5.3.9 differ in a fundamental way from the tuna models introduced previously.

Exercise 5.3.11

Consider an active management strategy for the tuna models in which the data consist of yearly effort and yields and the controls consist of harvest quotas. How could the quotas be used to help in model selection?

Bibliographic Notes

Rothschild and Suda (1977) provide a good discussion of the population dynamics of tuna. In the same book, Rothschild (1977) gives a discussion of fishing effort, its measurement, and quantification. Clark and Mangel (1979) provide other interpretations of the models in this section. Mangel (1982a) extends the models to include stochastic versions by replacing (5.3.14) and (5.3.15) with stochastic differential equations. Bifurcation theory [e.g., Arnold (1972)] became quite popular many years ago, and was known as "catastrophe theory" (Thom, 1975). The paper of Arnold, although highly technical, has good introductory sections. Cushing (1981) discusses aggregation of other species of fish. Sorensen (1980) provides a good introduction to nonlinear parameter estimation. The ideas of adaptive control in general are described by Bar-Shalom and Tse (1976). In particular, Wenk and Bar-Shalom (1980) and Baram and Sandell (1978) discuss the problem of model selection. The papers of Walters (1981), Walters and Hilborn (1976), and Walters and Ludwig (1982) describe applications of adaptive control in fishing systems. Young (1978) suggests a different approach for modeling complex ecological systems.

5.4 SEARCH EFFORT AND ESTIMATES OF STOCK SIZE

One way of assessing the size of fish stocks is through surveys such as those discussed in Section 5.2. It is not always feasible to conduct stock surveys, so alternate methods for estimating stock size should be investigated. One alternative is to use the catch–effort information generated by fisherman. The catch per unit effort model does this by assuming that catch is given by $Y = qEN$, where Y is the catch, N the population, E the effort, and q a parameter. Then, in the steady state, Y/E is proportional to N. A different approach can be based on the observation that fishermen must find fish before harvesting them and the rate at which fish are found may indicate stock size. Consequently, there may be a way of estimating stock size using encounter rate data. In this section the problem of estimating stock abundance based on search encounter rates is considered. Two models are used. In the first model depletion of the stock through harvesting is ignored; in the second model depletion is taken into account.

The first model is based on the following assumptions.

(1) The schools of fish move as cohesive units in the open ocean. Their distribution is unknown and there is no interaction between schools.

(2) The fishing boats search for fish and $1/\lambda$ is the mean time between sightings of schools. The parameter λ depends upon operational aspects (e.g., speed, detection width, size and density of schools) but is assumed to be independent of school size (which seems to be true at least for the tuna fishery).

(3) There is a probability p that a false detection of a school occurs. If a false detection occurs, the boat stops searching for an amount of time τ_1. That is, τ_1 is the time required to verify that the detection is actually a school. After this time, search resumes.

5.4. SEARCH EFFORT AND ESTIMATES OF STOCK SIZE

(4) With probability $1 - p$, a school is detected and the boat stops searching for τ hours. Here $\tau = \tau_1 + \tau_h$, with τ_1 the verification time and τ_h the harvest time.

(5) The random search model (Koopman, 1980) is used. According to this model the detections of schools are exponentially distributed with parameter $\lambda = Wv/A$. Here W is the sweep width of the searcher, v the velocity of the searcher, and the schools are distributed with one school every A (nautical miles)^2.

For this model one would be interested in the following kinds of questions: (1) What is the harvest (i.e., encounter) rate as a function of p, λ, τ_1, and τ_h? (2) Given that N schools were found in t hours, how does one estimate λ?

The process search–fish–search $\cdots$ is a renewal process (Karlin and Taylor, 1977). Consider a long time interval $[0, t]$ such that $t \gg 1/\lambda$ and $t \gg \tau$. Suppose that μ and σ^2 are the mean and variance in this renewal process and let $N(t)$ be the number of schools encountered in $[0, t]$. If $M(t)$ is the expected value of $N(t)$, then according to the renewal theorem $M(t)$ satisfies the equation

$$M(t) = \int_0^t M(t - y)\, dF(y) + F(t). \tag{5.4.1}$$

In this equation $F(t)$ is the distribution function of the first search–fish combination, with mean μ and variance σ^2; $M(t)$ is called the renewal function.

In order to verify (5.4.1), assume that the first encounter with a school of fish occurs at some time y. Then

$$E\{N(t)|\text{first event at } y\} = \begin{cases} 0, & \text{if } y > t, \\ 1 + M(t - y), & \text{if } y \leq t. \end{cases}$$

Averaging over all possible y gives

$$M(t) = E_y\{E\{N(t)|\text{first event at } y\}\}$$
$$= \int_0^t [1 + M(t - y)]\, dF(y)$$
$$= F(t) + \int_0^t M(t - y)\, dF(y),$$

which is the same as (5.4.1).

In order to apply (5.4.1), one needs to be able to compute $dF(t)$. For the assumptions described here, $F(t)$ can be explicitly computed. To do this, let $H(s)$ be the Heavyside function defined by $H(s) = 0$ for $s < 0$, $H(s) = 1$ for $s > 0$.

Exercise 5.4.1

Show that

$$F(t) = pH(t - \tau_1)\{1 - \exp[-\lambda(t - \tau_1)]\}$$
$$+ (1 - p)H(t - \tau)[1 - \exp[-\lambda(t - \tau)]\}. \qquad (5.4.2)$$

The density $dF(t)$ is found by differentiating (5.4.2). This is easy to do when one recognizes that $dH/ds = \delta(s)$, the Dirac delta function (see Chapter 1). Consider the case in which $t \gg \tau, \tau_1$. Evaluating $dF(t)$ and substituting it into (5.4.1) leads to the renewal equation

$$M(t) = p\{1 - \exp[-\lambda(t - \tau_1)]\} + (1 - p)\{1 - \exp[-\lambda(t - \tau)]\}$$
$$+ p\lambda \int_{\tau_1}^{t} \exp[-\lambda(y - \tau_1)] M(t - y) \, dy$$
$$+ (1 - p)\lambda \int_{\tau}^{t} \exp[-\lambda(y - \tau)] M(t - y) \, dy. \qquad (5.4.3)$$

Equation (5.4.3) is valid only if $t \gg \tau_1, \tau$. One way to solve (5.4.3) is to seek a formal asymptotic solution. To do this, assume that $M(t)$ has the form

$$M(t) = c_1 t + c_2 + c_3/t + O(1/t^2). \qquad (5.4.4)$$

One proceeds as follows. After substituting (5.4.4) into (5.4.3) and evaluating the integrals, terms are collected according to powers of t. Then the coefficient of each power of t is set equal to 0.

Exercise 5.4.2

Follow this procedure and show that it leads to

$$c_1 = \frac{\lambda}{1 + \lambda(p\tau_1 + (1 - p)\tau)}. \qquad (5.4.5)$$

Next, consider the general renewal equation

$$M(t) = F(t) + \int_0^t M(t - y) \, dF(y).$$

5.4. SEARCH EFFORT AND ESTIMATES OF STOCK SIZE

Use the ansatz (5.4.4) to obtain the general form of the solution, valid to order 1 (that is, compute c_1 and c_2).

Hence, the asymptotic form of the renewal function in (5.4.3) is

$$M(t) \simeq \frac{\lambda t}{1 + \lambda[p\tau_1 + (1-p)\tau]} + O(1) \tag{5.4.6}$$

Equation (5.4.6) shows clearly how the parameters of search, false detections, and harvest time enter into the rate of encounters of ships and schools [this rate is $M(t)/t$]. If no false detections occur, then

$$M(t) \simeq \lambda t/(1 + \lambda \tau). \tag{5.4.7}$$

Since $1/\lambda$ is a measure of time, it is useful to divide (5.4.6) by λ, giving

$$M(t) \simeq \frac{t}{1/\lambda + [p\tau_1 + (1-p)\tau]}. \tag{5.4.8}$$

The denominator in (5.4.8) is the average time spent per detection, so that (5.4.8) is a result that could almost be anticipated intuitively.

Exercise 5.4.3

Justify the assertion following (5.4.8).

When one computes the second term in the expansion (that is, c_2) for $p = 0$, the renewal function is given by

$$M(t) = 1/\mu + (\sigma^2 - \mu^2)/2\mu^2 + O(1/t) \tag{5.4.9}$$

$$= \lambda t/(\lambda\tau + 1) - [(\lambda\tau + 1)^2 - 1]/2(\lambda\tau + 1)^2 + O(1/t). \tag{5.4.10}$$

If a set time of τ hours results in an harvest of $H(\tau)$, then the long-run rate of harvest, for the case in which there are no false detections, is given by

$$R(t, \tau) = H(\tau)M(t)/t \tag{5.4.11}$$

$$= H(\tau)\left\{\frac{\lambda}{\lambda\tau + 1} - \frac{1}{t}\left(\frac{(\lambda\tau + 1)^2 - 1}{(\lambda\tau + 1)^2}\right) + O\left(\frac{1}{t^2}\right)\right\}. \tag{5.4.12}$$

Equations (5.4.10) and (5.4.12) can be used to draw some inferences about catch and effort. If effort is measured in terms of the time t at sea, then catch is roughly proportional to effort. But effort could also be measured in terms of search effectiveness $1/\lambda$ or set time τ, in which case catch is not proportional to effort.

Exercise 5.4.4

Compute the second term in the asymptotic expansion for the case when $p \neq 0$.

These results can be used to provide an estimate of the density of schools. For the model formulated here, this means providing an estimate of A in the formula $\lambda = Wv/A$, since v is presumed known and W can be estimated from other sources. If W and v are known, then A can be estimated once λ is known. A maximum likelihood estimate of λ can be obtained if one knows the distribution of $N(t)$. This distribution is given approximately by the following central limit theorem for renewal processes (Karlin and Taylor, 1977).

Let μ and σ^2 be the mean and variance of the first renewal and let $N(t)$ be the number of schools encountered in $[0, t]$. Then the quantity $[N(t) - (t/\mu)]/(t\sigma^2/\mu^3)^{1/2}$ is asymptotically (for $t \to \infty$) normally distributed with mean 0 and variance 1.

The mean and variance are found from the distribution (5.4.2). One finds after some calculations that

$$\mu = 1/\lambda + p\tau_1 + (1-p)\tau,$$
$$\sigma^2 = 1/\lambda^2 + p\tau_1^2 + (1-p)\tau^2 - [p\tau_1 + (1-p)\tau]^2. \tag{5.4.13}$$

Exercise 5.4.5

Verify equations (5.4.13).

Note that a τ, τ_1 cross-term appears in the variance σ^2. Also note that if $p \to 0$ or $p \to 1$, then $\sigma^2 \to 1/\lambda^2$. This is reasonable since then the only source of randomness is the search process (the time spent not searching is then purely deterministic). Rewrite (5.4.13) as

$$\mu = A/Wv + g_1(p, \tau, \tau_1),$$
$$\sigma^2 = A^2/(Wv)^2 + g_2(p, \tau, \tau_1). \tag{5.4.14}$$

Suppose that in an operation time of t that N schools are discovered and that the values of p, τ, τ_1, W, and v are known. The likelihood of observing N schools, $\mathscr{L}(N)$, is

$$\mathscr{L}(N) = \left[\frac{2\pi Wvt[A^2 + g_2(p, \tau, \tau_1)W^2v^2]}{[A + g_1(p, \tau, \tau_1)Wv]^3}\right]^{-1/2}$$
$$\times \exp\left[-\frac{1}{2}\left(N - \frac{tWv}{A + g_1(p, \tau, \tau_1)Wv}\right)^2\right.$$
$$\left.\times \left(\frac{tWv(A^2 + g_2(p, \tau, \tau_1)W^2v^2)}{[A + g_1(p, \tau, \tau_1)Wv]^3}\right)^{-1}\right]. \tag{5.4.15}$$

5.4. SEARCH EFFORT AND ESTIMATES OF STOCK SIZE

Table 5.4.1

MLE and Simplistic Estimates for λ

N	Estimate of λ	
	MLE (λ)	Simplistic (λ_s)
10	0.021	0.0198
20	0.047	0.0396
30	0.078	0.0595
40	0.116	0.0794
50	0.165	0.0992
60	0.228	0.1190
70	0.314	0.1389
80	0.440	0.1587
90	0.629	0.1786
100	0.968	0.1934

Equation (5.4.15) can be used to find the maximum likelihood estimate of λ by setting $\partial \mathscr{L}/\partial \lambda = 0$.

In the special case that $p = 0$, $\partial \mathscr{L}/\partial \lambda = 0$ reduces to $(\lambda = Wv/A)$ ·

$$0 = \frac{\partial}{\partial \lambda} \left\{ \frac{3}{2} \log(\lambda \tau + 1) - \frac{1}{2} \log \lambda - \frac{[N - \lambda t/(\lambda \tau + 1)]^2}{2 \lambda t} (\lambda \tau + 1)^3 \right\}. \quad (5.4.16)$$

Table 5.4.1 shows the MLE for λ as a function of N for $t = 504$ hours and $\tau = 4$ hours. In this case the maximum possible number of schools encountered is $t/\tau = 126$. If harvest time were not included, then a simplistic estimate of λ, denoted by λ_s, is $\lambda_s = N/t$. This is also shown in Table 5.4.1. The source of the difference between the MLE and simplistic estimates is the time spent fishing. The difference can be considerable even if the number of schools encountered is small (even for $N = 40$ the difference is more than 50%). A third possible estimate is the "semisimplistic" one, defined by

$$\lambda_{ss} = N/(t - N\tau). \quad (5.4.17)$$

It is remarkably accurate, for example, for $N = 10$, $\lambda_{ss} = 0.022$ and for $N = 100$, $\lambda_{ss} = 0.962$. These results suggest that the likelihood function is sharply peaked.

Exercise 5.4.6

What properties of the model would allow the estimate given by (5.4.17) to work well?

Exercise 5.4.7

Equations (5.4.15) and (5.4.17) are derived presuming that p, the probability of a false detection, is zero. Assuming that p is known precisely, derive analogs of (5.4.15–5.4.17). In many cases p will have to be estimated from the data as well. How would the procedure in (5.4.15–5.4.17) have to be modified?

If the time period of interest is long enough or if there are many vessels on the fishing ground, then depletion of the stock must be taken into account. In order to do this, the fundamental model must be changed. Assume that N vessels participate in the fishery. Instead of the Poisson assumption (which is equivalent to the first model discussed in this section)

$$\Pr\{\text{a vessel encounters another school in } (t, t+dt)\} = N\lambda\, dt + o(dt), \quad (5.4.18)$$

use the search-with-depletion assumption

$$\Pr\{\text{a vessel encounters another school in } (t, t+dt) | n \text{ caught up to } t\}$$
$$= N(\lambda - n\delta)\, dt + o(dt). \quad (5.4.19)$$

Here λ is proportional to the initial density of fish schools and δ is proportional to the reduction in density when a school is harvested. Harvest time will be ignored in the rest of this section.

Exercise 5.4.8

Suppose that the initial number of schools is known to be n_0 and that the area of the entire region is A. Show that the assumption of random search implies that $\lambda = n_0 Wv/A$ and $\delta = Wv/A$.

The difficulty in analyzing the data is that n_0 is not known: it must be estimated from the data which consist of the pair (n, T), where n is the number of schools found with search time T. Two estimation methods will now be described. The first provides an estimate $\hat{\lambda}$ for λ from which an estimate $\hat{n}_0$ is found by setting $\hat{n}_0 = \hat{\lambda}/\delta$. The second method provides a direct estimate for n_0.

Equation (5.4.19) implies that the time to find the nth school is exponentially distributed with parameter $N(\lambda - (n-1)\delta)$. Assume that a data set $\{T_1, \ldots, T_n\}$ is available at time $T = \sum_{i=1}^{n} T_i$, where T_i is the time spent locating the ith school. Assuming that the T_i are independent random variables, the likelihood of this data set is

$$\mathscr{L}_n = \prod_{i=1}^{n} N(\lambda - (i-1)\delta) \exp\{-NT_i(\lambda - (i-1)\delta)\} \quad (5.4.20)$$

5.4. SEARCH EFFORT AND ESTIMATES OF STOCK SIZE

and the log-likelihood is

$$\mathscr{L}_n = \sum_{i=1}^{n} \log(N(\lambda - (i-1)\delta)) - NT_i(\lambda - (i-1)\delta). \quad (5.4.21)$$

Setting $\partial \mathscr{L}_n/\partial \lambda = 0$ gives an equation for the maximum likelihood estimate of λ. This equation is

$$\sum_{i=1}^{n} \frac{1}{\lambda - (i-1)\delta} = N \sum_{i=1}^{n} T_i = NT. \quad (5.4.22)$$

Observe from (5.4.22) that the estimate of λ depends only upon the total search time T and not on any of the intermediate times T_i.

Let $\hat{\lambda}$ be the solution of (5.4.22). To estimate its variance, one can again use the Fisher information. From (5.4.21)

$$\frac{\partial}{\partial \lambda} \mathscr{L}_n = \sum_{i=1}^{n} \frac{1}{\lambda - (i-1)\delta} - NT_i,$$

so that

$$\frac{\partial^2}{\partial \lambda^2} \mathscr{L}_n = -\sum_{i=1}^{n} \frac{1}{[\lambda - (i-1)\delta]^2}. \quad (5.4.23)$$

Equation (5.4.23) clearly shows that the estimate given by (5.4.22) will be MLE.

The Fisher information number is then

$$I(\lambda) = -\mathrm{E}\{\partial^2 \mathscr{L}_n/\partial \lambda^2\}|_{\hat{\lambda}}$$

$$= \int \sum_{i=1}^{n} \frac{1}{[\lambda - (i-1)\delta]^2} \prod \lambda_i \exp(-\lambda_i T_i) \, dT_i \bigg|_{\hat{\lambda}}, \quad (5.4.24)$$

where $\lambda_i = \lambda - (i-1)\delta$. The integral in (5.4.24) is easily done, giving

$$I(\hat{\lambda}) = \sum_{i=1}^{n} \frac{1}{[\hat{\lambda} - (i-1)\delta]^2}. \quad (5.4.25)$$

In the Bayesian interpretation, $\lambda \simeq N(\hat{\lambda}, 1/nI(\hat{\lambda}))$. This interpretation must be modified slightly, since it allows λ to take on negative values. To avoid this difficulty, condition on $\lambda > 0$ and thus write the following for the probability density $f(\lambda, n)$:

$$f(\lambda, n) = \frac{\exp[-(\lambda - \hat{\lambda})^2 nI(\hat{\lambda})/2]}{\int_0^\infty \exp[-(\lambda - \hat{\lambda})^2 nI(\hat{\lambda})/2] \, d\lambda}, \quad \lambda \geq 0. \quad (5.4.26)$$

Exercise 5.4.9

Obtain approximate expressions for the normalization constant in (5.4.26) for the limiting cases $\hat{\lambda} \simeq 0$ and $\hat{\lambda} \gg 0$.

Given a catch and search time (n, T), one obtains an estimate for λ. With this estimate, $\hat{\lambda}$, one can make predictions about future catches. Observe first that the estimate for the number of schools is now $(\hat{\lambda} - n\delta)/\delta$. Suppose that there are an additional T_s units of search expended. Let $H(T_s)$ denote the number of schools found with search effort T_s. The conditional moments of $H(T_s)$ are

$$E\{H(T_s)|\lambda\} = [(\lambda - n\delta)/\delta][1 - \exp(-N\delta T_s)],$$
$$\text{Var}\{H(T_s)|\lambda\} = [(\lambda - n\delta)/\delta][1 - \exp(-N\delta T_s)]\exp(-N\delta T_s). \quad (5.4.27)$$

One is really interested in the unconditional version of (5.4.27), obtained by averaging over λ. This gives

$$E\{H(T_s)\} = \int_{n\delta}^{\infty} [(\lambda - n\delta)/\delta][1 - \exp(-N\delta T_s)] f(\lambda, n) \, d\lambda, \quad (5.4.28)$$

so that

$$E\{H(T_s)\} = \frac{1}{\delta} \left\{ \frac{[1 - \exp(-N\delta T_s)]}{\int_0^\infty \exp[-(\lambda - \hat{\lambda})^2 nI(\hat{\lambda})/2] \, d\lambda} \right\}$$
$$\times \int_0^\infty u \exp[-(u + n\delta - \hat{\lambda})^2 nI(\hat{\lambda})/2] \, du. \quad (5.4.29)$$

Exercise 5.4.10

Verify equations (5.4.27). [*Hint*: See (5.4.30) and Chapter 1.] How can the integral in (5.4.28) be dealt with?

The calculation just done assumes that the observation time T is fixed and that the number of schools discovered n varies. The second method of estimation, in which n is fixed and T varies, is the following. Recall that in Chapter 1 it was shown that (5.4.19) leads to the binomial formula

$$\Pr\{N \text{ vessels finding } n \text{ schools in } (0, T)\}$$
$$= \binom{\lambda/\delta}{n}[1 - \exp(-N\delta T)]^n \exp[-N\delta T(\lambda/\delta - n)]. \quad (5.4.30)$$

Equation (5.4.30) is the starting point for the next calculation.

Assume that n schools were found in $(0, T)$ by the N vessels. The question one then asks is, What can be inferred about λ/δ? For ease of calculation, set

$$q = e^{-N\delta T}, \qquad n_0 = \lambda/\delta. \quad (5.4.31)$$

5.4. SEARCH EFFORT AND ESTIMATES OF STOCK SIZE

Equation (5.4.30) can be viewed as a likelihood function for n_0 given n. In particular, set

$$L(n_0; n) = \binom{n_0}{n} p^n q^{(n_0 - n)}. \qquad (5.4.32)$$

Here $p = 1 - q$. In order to find the maximum likelihood estimate for n_0, n_0^*, begin by comparing $L(n_0 + 1; n)$ and $L(n_0; n)$. This gives

$$\frac{L(n_0 + 1; n)}{L(n_0; n)} = \frac{\binom{n_0+1}{n} p^n q^{n_0+1-n}}{\binom{n_0}{n} p^n q^{n_0-n}} = \frac{\binom{n_0+1}{n}}{\binom{n_0}{n}} q = \left(\frac{n_0 + 1}{n_0 + 1 - n}\right) q. \qquad (5.4.33)$$

The likelihood ratio in (5.4.33) is 1 if

$$[(n_0 + 1)/(n_0 + 1 - n)]q = 1. \qquad (5.4.34)$$

Solving this equation for n_0 gives

$$n_0 = n/(1 - q) - 1. \qquad (5.4.35)$$

Since n_0 should be an integer, choose the following as the estimator for n_0:

$$\hat{n}_0 = [n/(1 - q)], \qquad (5.4.36)$$

where $[x]$ is the integer part of x.

Exercise 5.4.11

What should be done if $n/(1 - q)$ is actually an integer?

Exercise 5.4.12

Show that (5.4.36) and (5.4.22) lead to virtually the same estimates for n_0 (or λ).

The distribution of $\hat{n}_0$ is not as easily calculated, but (5.4.32) is very useful if one adopts a Bayesian approach. For example, suppose that $g(n_0)$ is a prior distribution on n_0 and assume that n schools were found in time T. If $g(n_0|n)$ is the posterior distribution for n_0 given n, then

$$g(n_0|n) = L(n_0; n)g(n_0) \Big/ \sum_{n_0=n}^{\infty} L(n_0; n)g(n_0) \qquad (5.4.37)$$

A particularly simple result is obtained if one uses the improper prior distribution given by

$$g(n_0) = 1 \quad \text{for all} \quad n_0. \qquad (5.4.38)$$

Then (5.4.37) becomes

$$g(n_0|n) = L(n_0; n) \Big/ \sum_{n_0=n}^{\infty} L(n_0; n). \qquad (5.4.39)$$

Exercise 5.4.13

Verify the identity, valid for $|x| < 1$,

$$\sum_{N=n}^{\infty} \binom{N}{n} x^{N-n} = (1-x)^{-n-1}. \tag{5.4.40}$$

Using the identity (5.4.40) in the denominator of (5.4.39) gives

$$\sum_{n_0=n}^{\infty} L(n_0; n) = \sum_{n_0=n}^{\infty} \binom{n_0}{n} p^n q^{n_0-n}$$

$$= p^n \sum_{n_0=n}^{\infty} \binom{n_0}{n} q^{n_0-n} = p^n (1-q)^{-n-1}. \tag{5.4.41}$$

Since $p = 1 - q$, the denominator in (5.4.39) is simply p^{-1}. Consequently, the posterior distribution for n_0 is

$$g(n_0 | n) = \binom{n_0}{n} p^{n+1} q^{n_0-n} \tag{5.4.42}$$

with $n_0 = n, n+1, \ldots$. Equations (5.4.36) and (5.4.42) characterize the stock information provided by search effort.

Exercise 5.4.14

A problem sometimes faced by fishery managers is assigning stock estimates to regions where some effort was expended but no fish were caught. Show how the results just derived can be used to help answer this question. Instead of the improper prior distribution (5.4.38), use the prior distribution

$$g(n_0) = 1 \quad \text{on} \quad n_0 = 0, 1, 2, \ldots, N_m, \tag{5.4.43}$$

where N_m is a (given) maximum possible number of schools in the region (how could it be estimated?) [*Hint*: recall that $\sum_{n=0}^{N} (1-p)^n = (1 - (1-p)^{N+1})/p$]. What other prior distributions might be reasonable? What posterior distribution does one want to compute?

Bibliographic Notes

The first part of this section comes from Mangel (1982b). Renewal theory and the central limit theorem are discussed in many books on stochastic processes [e.g., Karlin and Taylor (1977), Feller (1971), and Ross (1970)]. A good introduction to asymptotic expansions is provided by Lin and Segel (1974). Much of the latter part of this section is taken from Mangel and Beder

5.4 SEARCH EFFORT AND ESTIMATES OF STOCK SIZE

(1984), who show how these results can be used to determine the length of fishing seasons, to determine confidence intervals for n_0, and apply to the fishery for Pacific Ocean Perch. Improper prior distributions are discussed by DeGroot (1970) and Berger (1980). The random variables T_i in this section are interesting because they are independent but not identically distributed. The book by Barlow, Bartholomew, Bremner, and Brunk (1972) provides statistical techniques for dealing with this kind of data. Useful combinatoric identities, similar to (5.4.40), are found in Riordan (1979).

6

Management of Fluctuating Renewable Resource Stocks

In this chapter the optimal use of uncertain renewable resource stocks is considered. Such problems are quite difficult; the techniques and discussion presented here can be viewed as an introduction to this fascinating problem area in resource management.

The chapter begins with the question faced by a small coastal nation: should it build a big fishing fleet or sell rights to fish in its exclusive economic zone (EEZ)? Then the question of optimal management of a fluctuating stock is considered. Cases of known and unknown stock dynamics are treated. Next the market dynamics for uncertain resource stocks are studied. In the last section strategies for resource management that minimize the chance of undesirable events are discussed.

When dealing with uncertainty and noise, flexibility and adaptability are important aspects of a control policy. Walters and Hilborn (1976) introduced the concept of adaptive control in fishing systems.

6.1 THE ALLOCATION OF FISHING RIGHTS

The model used in this section comes from Beddington and Clark (1984). Assume that a nation is developing a new fishery for a single stock. The questions faced by the nation are: (1) How big a fleet should be developed?

6.1. THE ALLOCATION OF FISHING RIGHTS

Table 6.1.1

Choices for the Stock–Recruitment Formula

Choice	Name
$F(S, w) = R(w)$	Uncoupled random generations
$ = w_1 S + R(w_2)$	Stochastic recruitment and survival (w_1 and w_2 are independent random variables)
$ = S e^{\alpha + bS + w}$	Stochastic Ricker curve

(2) Should fishing rights be sold to foreigners? (3) If so, at what level?

The biological model relating biomass in year $k+1$, B_{k+1}, and remaining stock in year k, S_k, is assumed to take the form

$$B_{k+1} = F(S_k, w_k), \tag{6.1.1}$$

where w_k is a noise term. It is assumed that the w_k are independent, identically distributed random variables. Some choices for $F(\cdot, \cdot)$ are shown in Table 6.1.1. The first choice would be most appropriate for stocks similar to shrimp or prawns; the third choice would be best for stocks similar to salmon.

Assume that there are two fleets: a domestic fleet (denoted by superscript d) and a foreign fleet (denoted by superscript f). The catches by the domestic and foreign fleets in year k are denoted respectively by H_k^d and H_k^f. Thus,

$$S_k = B_k - (H_k^f + H_k^d);$$

sometimes the notation $H_k = H_k^f + H_k^d$ will be used. Assume that p^d is the domestic unit price of fish and that p^f is the unit price that the foreigners pay in order to fish. The problem is most interesting if $p^d > p^f$. If $C^d(B)$ is the cost of harvesting a unit of fish from a stock at level B, then the domestic net profit in year k is given by

$$\pi_k = \int_{B_k - H_k^d}^{B_k} [p^d - C^d(x)] \, dx + p^f H_k^f. \tag{6.1.2}$$

The problem is to determine the sequence of quotas $\{H_k^d, H_k^f\}$ and the domestic harvest capacity K to maximize the objective functional

$$J = E\left\{\sum_{k=0}^{\infty} \left(\frac{1}{1+\alpha}\right)^k \pi_k\right\} - c_1(K), \tag{6.1.3}$$

where $c_1(K)$ is the cost of a domestic capacity of size K and $1/1 + \alpha$ is a discounting factor.

To add a certain level of biological realism to the problem, add to (6.1.3) a constraint on the escapement (the stock size at the end year). This constraint is that $S_k \geq B^*$. Here B^* is the escapement level and $S_k = B_k - H_k - H_k^f$ is the remaining stock at the end of the season.

Exercise 6.1.1

Why is it reasonable to add the constraint on escapement?

To begin, consider the situation in which stock levels are uncoupled. Then the return in each year is completely independent of previous years' returns. Consequently, one only needs to evaluate

$$E\left\{\max_{H^d, H^f} \int_{B-H^d}^{B} \varphi(x)\,dx + p^f H^f\right\} \quad \text{such that} \quad H^d \leq K, \quad (6.1.4)$$

where $\varphi(x) = p^d - C^d(x)$. Define a biomass level B^∞ by

$$C^d(B^\infty) = p^d - p^f. \quad (6.1.5)$$

At biomass B^∞ the foreign and domestic revenues are the same. If B falls below B^∞, then foreign revenue exceeds domestic revenue [assuming that $C^d(x)$ increases as x decreases].

With these assumptions one can reason as follows. The total harvest must leave a stock level B^*. Consider the domestic catch first. If the current biomass is B, the domestic fleet should take either $B - B^\infty$ (reducing the stock to the level of unprofitable fishing) or the capacity K. The foreigners should be allowed to take the rest of the stock. Consequently, the appropriate harvest quotas are (assuming that $B^\infty > B^*$)

$$H^{d*} = \min(B - B^\infty, K),$$
$$H^{f*} = \max(B - K - B^*, 0) + \max(B^\infty - B^*, 0). \quad (6.1.6)$$

The optimal quota for the foreign fleet is composed of two terms. The first term represents surplus capacity provided by the foreign fleet; the second term represents the uneconomic surplus.

To determine the optimal value of K, one picks K to maximize the revenue obtained over a single season. To find this optimal value, proceed as follows. Let $\pi(B, K)$ denote the value of the revenue function in (6.1.14) when the biomass level is B, the capacity is K, and the optimal quotas (6.1.6) are employed. The expected value of $\pi(B, K)$ is then

$$E\{\pi(B, K)\} = \int \pi(x, K)f(x)\,dx,$$

6.1. THE ALLOCATION OF FISHING RIGHTS

where $f(x)$ is the probability density for the biomass. Setting $\hat{\alpha} = 1/(1 + \alpha)$, (6.1.3) becomes

$$J = E\{\pi(B, K)\} \sum_{k=0}^{\infty} \hat{\alpha}^k - c_1(K)$$

$$= E\{\pi(B, K)\}(1 - \hat{\alpha})^{-1} - c_1(K). \quad (6.1.7)$$

The optimal value of K is that which maximizes the right-hand side of (6.1.7).

Exercise 6.1.2

Suppose that $C^d(B) = c_0/B$, $\log B \sim N(\mu, \sigma^2)$, and $c_1(K) = \alpha K$. Characterize the optimal value of K.

In the general case of coupled stocks, the intraseasonal quotas H^d and H^f can be found from (6.1.6). For coupled stocks, however, the escapement B^* must be treated as a parameter to be used in the optimization. There are now two cases to consider.

Case 1: $B^\infty < B^*$. In this case reasoning similar to the reasoning that lead to (6.1.6) leads to

$$H^d(B) = \begin{cases} 0, & B \leq B^*, \\ B - B^*, & B^* < B \leq B^* + K, \\ K, & B^* + K < B; \end{cases}$$

$$H^f(B) = \begin{cases} 0, & B \leq B^* + K, \\ B - (B^* + K), & B^* + K < B. \end{cases}$$

Case 2: $B^\infty > B^*$. In this case reasoning similar to the reasoning that lead to (6.1.6) leads to

$$H^d(B) = \begin{cases} 0, & B \leq B^\infty, \\ B - B^\infty, & B^\infty < B \leq B^\infty + K, \\ K, & B^\infty + K < B; \end{cases}$$

$$H^f(B) = \begin{cases} 0, & B \leq B^*, \\ B - B^*, & B^* < B \leq B^\infty, \\ B^\infty - B^*, & B^\infty < B \leq B^\infty + K, \\ B - (B^* + K), & B^\infty + K < B. \end{cases}$$

Exercise 6.1.3

Verify the results given above for cases 1 and 2.

In order to determine B^*, assume that $B_k > B^*$ with probability 1. The net profit in year k is now

$$\pi(B_k; B^*, K) = \int_{B_k - H^d(B_k)}^{B_k} \varphi(x)\, dx + p^f H^f(B_k). \tag{6.1.8}$$

Setting $B_k = F(B^*, w)$, as in (6.1.1), the optimal value of B^* is determined by maximizing $E = \{\pi(F(B^*, w); B^*, K)\}$. Once the optimal value of B^* is known, the optimal value of K is determined as in (6.1.7).

Exercise 6.1.4

What should be done if $\Pr\{B_k < B^*\} > 0$?

The results obtained thus far are based on the assumption that the stock levels B_k are accurately known. It may be more realistic to assume that only the probability distribution for B_k is known. Then the management strategy should involve *effort* rather than catch quotas. If E^d and E^f are the domestic and foreign effort quotas, then a reasonable intraseasonal model for catch is (assuming the domestic fishermen fish first)

$$H^d = (1 - e^{-qE^d}) B_k, \tag{6.1.9}$$

$$H^f = e^{-qE^d}(1 - e^{-qE^f}) B_k, \tag{6.1.10}$$

where q is the catchability coefficient and B_k the stock level at the start of year k.

Exercise 6.1.5

What model is used to derive the formula $H = (1 - e^{-qE})B$?

Assuming that the return function from the foreign fleet is simply $p^f H^f$, the net return in year k is given by

$$\pi_k = p^d H^d - c^d E^d + p^f H^f \tag{6.1.11}$$

$$= [p^d(1 - e^{-qE^d}) + p^f e^{-qE^d}(1 - e^{-qE^f})] B_k - c^d E^d, \tag{6.1.12}$$

where c^d is the total domestic cost of effort.

For the case of uncoupled stocks, the expected revenue in year k is

$$E\left\{\left(\frac{1}{1+\alpha}\right)^k \pi_k\right\} = \frac{\langle B \rangle}{(1+\alpha)^k}[p^d(1 - e^{-qE^d}) + p^f e^{-qE^d}(1 - e^{-qE^f})] - \frac{c^d E^d}{(1+\alpha)^k}. \tag{6.1.13}$$

In this equation $\langle B \rangle$ denotes the average value of B. The total discounted profit is found by summing (6.1.13) over k. This sum is clearly similar to (6.1.13), with $1/(1+\alpha)^k$ replaced by $1/(1-\hat{\alpha})$. Thus, the total discounted profit is maximized by setting E^f as large as possible. Taking the derivative with respect to E^d and setting it equal to 0 gives the optimal level of domestic effort

$$E^{d*} = \frac{1}{q}\ln\left\{\frac{q\langle B\rangle}{c^d}[p^d - p^f(1 - e^{-qE^f})]\right\}. \qquad (6.1.14)$$

Exercise 6.1.6

What is the optimal behavior for the foreign fleet, that is, how much effort should it buy?

Exercise 6.1.7

What is the effort strategy for a coupled stock relationship?

Bibliographic Notes

Cushing (1981) discusses different biological models and which models are appropriate for which kind of stock. Agnello and Anderson (1981) discuss models for production responses in multispecies fisheries. The results presented in this section are quite simple because the stochastic dynamic programming equations simplify through the use of myopic approximations (Heyman and Sobel, 1984). Measurement errors in stock size are undoubtedly common, so one should try to take them into account. Such errors are discussed in Section 6.3.

6.2 HARVESTING A RANDOMLY FLUCTUATING POPULATION

The problem of optimally harvesting a randomly fluctuating population has received considerable attention in recent years [e.g., Anderson (1975), Gleit (1978), Hutchinson and Fischer (1979), Ludwig (1979a, b, 1980), and Lewis (1981)]. This section contains an introduction to some of these analyses.

The population of interest is characterized by a single biomass variable $X(t)$ that is assumed to satisfy the SDE

$$dX = \begin{cases} [f(X, t) - h(X, t)]\, dt + \sigma(X, t)\, dW, & X > 0, \\ 0, & X = 0. \end{cases} \quad (6.2.1)$$

In this equation $f(X, t)$ is the natural (deterministic) growth rate, $h(X, t)$ the harvest rate, $\sigma(X, t)$ a measure of the intensity of the noise, and $W(t)$ a standard Browian motion process with $E\{dW\} = 0$ and $E\{(dW)^2\} = dt$. The profit from a harvest rate $h(X, t)$ is assumed to be given by

$$\pi(X, t, h(X, t)) = [p(t) - c(X, h, t)]h(X, t), \quad (6.2.2)$$

where $p(t)$ is the price per unit harvest and $c(X, h, t)$ the cost of harvesting a unit biomass when the stock size is X and the harvest rate is h. Assume that $U(\cdot)$ is the social utility function associated with profit, $D(y, s)$ is a preservation value for a stock at level y at time s, and the discount rate is ρ. A reasonable objective functional is then

$$J = E\left\{ \int_0^T e^{-\rho t} U(\pi(X, t))\, dt + e^{-\rho T} D(X(T), T) \right\}. \quad (6.2.3)$$

The DPE for this problem is obtained by setting

$$J(x, t) = \max_h E\left\{ \int_t^T e^{-\rho s} U(\pi(X, s))\, ds + e^{-\rho T} D(X(T), T) \,|\, X(t) = x \right\}. \quad (6.2.4)$$

By standard methods the DPE for $J(x, t)$ is

$$0 = J_t + \max_h \{\tfrac{1}{2}\sigma^2(x, t) J_{xx} + [f(x, t) - h(x, t)] J_x + e^{-\rho t} U(\pi)\}. \quad (6.2.5)$$

Gleit (1978) studied the following case:

(1) $f(x, t) = k(t)x$,
(2) $U(\pi) = \pi^\gamma/\gamma$, $\quad 0 < \gamma < 1$,
(3) $D(x, t) = \alpha^{(1-\gamma)} x^\gamma/\gamma$, $\quad (6.2.6)$
(4) $c(X, t)h(X, t) = B(t)X(t)$,
(5) $\sigma(X, t) = \sigma(t)X$.

These assumptions are interpreted as follows.

(1) The population grows exponentially in the deterministic limit, namely, there are no density-dependent effects.
(2) The utility function indicates risk aversion.
(3) The preservation value indicates risk aversion.
(4) The cost of harvesting a unit biomass increases with population size.
(5) The fluctuations are environmental in origin.

6.2. HARVESTING A RANDOMLY FLUCTUATING POPULATION

Exercise 6.2.1

Criticize these assumptions. For what kind of operational situation do they apply?

Assuming an internal maximum and maximizing (6.2.5) over h gives the equation

$$J_x = e^{-\rho t} p(t) U'(\pi) \tag{6.2.7}$$

for the optimal value of h. In light of (6.2.6) the optimal harvest is

$$h^* = \frac{1}{p(t)} \left[\frac{e^{\rho t} J_x}{p(t)} \right]^{1/\gamma - 1} + \frac{B(t)x}{p(t)}. \tag{6.2.8}$$

Once J_x is known, (6.2.8) gives the optimal feedback harvest rate. Substituting back into the DPE and using (6.2.6) gives the nonlinear equation

$$0 = J_t + \frac{1}{2}\sigma(t)^2 x^2 J_{xx} + \left\{ k(t) - \frac{B(t)}{p(t)} \right\} x J_x + e^{-\rho t} \left(\frac{1-\gamma}{\gamma} \right) \left[\frac{e^{\rho t} J_x}{p(t)} \right]^{\gamma/\gamma - 1}. \tag{6.2.9a}$$

The end condition associated with (6.2.9a) is read off from the functional (6.2.3) and the assumptions (6.2.6). It is

$$J(x, T) = \alpha^{1-\gamma} e^{-\rho T} x^{\gamma}/\gamma. \tag{6.2.9b}$$

Equation (6.2.9a) is a nonlinear differential equation for $J(x, t)$. The results in Chapter 2 showed that sometimes the correct form of the solution can be guessed. In light of the end condition (6.2.9b), seek a solution of (6.2.9) in the form

$$J(x, t) = e^{-\rho t} H(t)^{1-\gamma} x^{\gamma}/\gamma, \tag{6.2.10}$$

where $H(t)$ is to be determined. Observe that (6.2.10) has the same form as the end condition (6.2.9b). Substituting (6.2.10) into (6.2.9a) gives an ordinary differential equation for $H(t)$. This equation is

$$H'(t) + \delta(t) H(t) + (1/p(t))^{\gamma/\gamma - 1} = 0, \tag{6.2.11a}$$

where

$$\delta(t) = \left[-\frac{\rho}{\gamma} + k(t) - \frac{\sigma^2(t)}{2}(1-\gamma) - \frac{B(t)}{p(t)} \right] \frac{\gamma}{1-\gamma}. \tag{6.2.11b}$$

The end condition is $H(T) = \alpha$.

Exercise 6.2.2

What conditions are needed on $\delta(t)$, $p(t)$, and γ to ensure that a solution of (6.2.11a, b) exists? Compute the solution of (6.2.11a).

Exercise 6.2.3

What can be said about $\partial H/\partial(\sigma^2)$? How is this derivative interpreted?

Once $H(t)$ is known, the optimal harvest rate is found from (6.2.7), (6.2.8), and (6.2.10). It is given by

$$h^*(x, t) = x\left[\frac{B(t)}{p(t)} + \frac{1}{H(t)[p(t)^{\gamma/\gamma - 1}]}\right], \qquad (6.2.12)$$

and the optimal value of the objective functional is

$$J(x_0, 0) = x_0^\gamma H(0)^{1-\gamma}/\gamma. \qquad (6.2.13)$$

According to (6.2.12) the optimal harvest rate increases with an increased cost/price ratio.

Exercise 6.2.4

What can be said about $\partial^2 J(x_0, 0)/\partial(\sigma^2)$ when $J(x_0, 0)$ is given by (6.2.13)? Why would the harvest rate increase with the cost/price ratio?

Exercise 6.2.5

As a specific case set $p(t) = p_0 e^{jt}$ and $B(t) = B_0 e^{jt}$, where $j > 0$ is a given inflation rate. Find $h^*(x, t)$ and $J(x_0, 0)$.

The results presented thus far are very model dependent. A more general analysis follows the work of Ludwig (1979a, b) and Ludwig and Varah (1979). By way of motivation, recall that often fish stock dynamics are described by stock–recruitment curves. That is, let R_{n+1} denote the stock size at the start of period $n + 1$ and let S_n denote the stock size at the end of period n. For example, the deterministic Ricker formula is

$$R_{n+1} = S_n \exp\{a + bS_n\}. \qquad (6.2.14)$$

A possible stochastic stock–recruitment curve is (see Table 6.1.1 for some other choices)

$$R_{n+1} = S_n \exp\{a + bS_n + \tilde{u}_n\}, \qquad (6.2.15)$$

where the $\tilde{u}_n$ are random variables. If $\tilde{u}_n$ is normal, then for fixed S_n, $\log R_{n+1}$ is a normally distributed random variable. This observation motivates working with the logarithm of stock size as the fundamental variable. Thus, define a new set of variables by

$$\begin{aligned} Y &= \log X, & \tilde{f}(Y) &= f(X, t)/X, \\ E &= h(X, t)/X, & \tilde{\sigma}(Y) &= \sigma(X, t)/X. \end{aligned} \qquad (6.2.16)$$

6.2. HARVESTING A RANDOMLY FLUCTUATING POPULATION

Exercise 6.2.6

Interpret the last three variables in (6.2.16).

The stock dynamics are assumed to be given by

$$dY = [\tilde{f}(Y) - E]\,dt + \sigma(Y)\,dW. \tag{6.2.17}$$

In this equation E will be called fishing effort.

Exercise 6.2.7

How are the equations for the stock dynamics (6.2.1) and (6.2.17) related?

A final transformation is done on the cost term. The revenue from harvesting at stock level X is $[p - C(X)]h(X, t)$. The transformation (6.2.16) suggests that one should also define a cost term $\tilde{C}(Y)$ such that

$$[p - C(X)]h(X, t) = [p - C(\tilde{Y})]Ee^Y. \tag{6.2.18}$$

Equation (6.2.18) is an implicit equation defining $C(\tilde{Y})$. For the revenue function set $n(Y) = e^Y[p - \tilde{C}(Y)]$; $n(Y)E$ is then a measure of revenue. The functional for the optimal control problem is taken to be

$$J(x, t) = \max_E \mathcal{E}\left\{ \int_t^\infty e^{-\delta s} n(Y(s))E\,ds \,\Big|\, Y(t) = x \right\}. \tag{6.2.19}$$

Here $\mathcal{E}$ denotes the expectation over paths of the process $Y(t)$. The DPE is

$$J_t + \tfrac{1}{2}\sigma^2(x)J_{xx} + \max_E\{J_x(\tilde{f}(x) - E) + e^{-\delta t}n(x)E\} = 0.$$

Setting $J(x, t) = e^{-\delta t}v(x)$ and substituting into the DPE gives

$$\tfrac{1}{2}\sigma^2(x)v_{xx} + \max_E\{v_x(\tilde{f}(x) - E) + n(x)E\} - \delta v = 0. \tag{6.2.20}$$

In this case a "bang-bang" control for E is optimal. This control is

$$E^* = \begin{cases} E_{\max} & \text{if } n(x) > v_x, \\ E_{\min} & \text{if } n(x) < v_x. \end{cases} \tag{6.2.21}$$

Here $E_{\max}$ and $E_{\min}$ are the exogenously given maximum and minimum values of effort. Dropping the tilde on f, E is understood to be E^* and we assume that $\tfrac{1}{2}\sigma^2(y) = \varepsilon D(y)$, where $0 < \varepsilon \ll 1$. The DPE is now

$$\varepsilon D v_{xx} + [f(x) - E]v_v - \delta v + n(x)E = 0. \tag{6.2.22}$$

Equation (6.2.22) is a singular perturbation problem, since if ε is set equal to 0 the order of the equation reduces. Before analyzing the equation, certain properties of $v(x)$ will be determined. Growth conditions associated with (6.2.22) can be determined as follows.

Integrate (6.2.22) once to obtain

$$\frac{d}{dx}\left\{v_x \exp\left[\int_0^x \frac{2[f(y) - E(y)]}{\varepsilon D(y)} dy\right]\right\}$$
$$= \frac{2}{\varepsilon D(x)} \exp\left[\int_0^x \frac{2[f(y) - E(y)]}{\varepsilon D(y)} dy\right][\delta v - n(x)E]. \quad (6.2.23)$$

As $y \to \infty$, $f(y)$ must be negative since it is the per capita growth rate. If $y \to -\infty$ (so that the population $\to 0$), f must go to zero. Hence if $\mu(x)$ is defined by

$$\mu(x) = \int_0^x \frac{2[f(y) - E(y)]}{\varepsilon D(y)} dy, \quad (6.2.24)$$

then $\mu(x) \to -\infty$ as $x \to -\infty$. The integrability of (6.2.23) implies the growth condition

$$v_x e^{\mu(x)} \to 0 \quad \text{as} \quad x \to \pm\infty. \quad (6.2.25)$$

The problem now is to understand the DPE (6.2.22). If ε in (6.2.22) is small, one can set it equal to zero and study the resulting equation. Setting $\varepsilon = 0$ in (6.2.22) shows that the DPE becomes

$$[f(x) - E]v_x - \delta v + n(x)E = 0. \quad (6.2.26)$$

A good first step is to analyze (6.2.26).

Exercise 6.2.8

To what deterministic problem does (6.2.26) correspond?

The problems (6.2.20) and (6.2.26) have bang-bang controls, except at switching points or curves for the effort level. The plan is to assume that the number of switches is known but that the locations of the switching points for effort are unknown.

Exercise 6.2.9

How are these switching points related to the singular paths in deterministic control theory?

To start assume that effort, given by (6.2.21), switches at only one point, say, y. Then set

$$E(x, y) = \begin{cases} E_{\max}, & x > y, \\ E_{\min}, & x < y, \\ f(y), & x = y. \end{cases} \quad (6.2.27)$$

6.2. HARVESTING A RANDOMLY FLUCTUATING POPULATION

Note that if (6.2.27) is followed and the level $x = y$ is hit, then $E(y, y) = f(y)$ and, deterministically, the stock level does not change.

Define $g(x, y) = f(x) - E(x, y)$. The goal is to determine y. Think of v in (6.2.26) as $v(x, y)$. Then the DPE can be rewritten as

$$g(x, y) \frac{\partial}{\partial x} v(x, y) - \delta v(x, y) + n(x)E(x, y) = 0. \quad (6.2.28)$$

At $x = y$, $g(y, y) = 0$ and (6.2.28) becomes

$$v(y, y) = f(y)n(y)/\delta. \quad (6.2.29)$$

Exercise 6.2.10

Show that $\partial v(x, y)/\partial x$ is continuous at y, that is, that

$$\lim_{x \downarrow y} \frac{\partial}{\partial x} v(x, y) = \lim_{x \downarrow y} \frac{\partial}{\partial x} v(x, y) = n(y). \quad (6.2.30)$$

If both of these limits exist, then the mixed derivative $\partial^2 v/\partial x \, \partial y$ must also exist.

The switching point y can be found by choosing it to maximize $v(x, y)$. Thus the optimal switching point is determined by the condition

$$\frac{\partial}{\partial y} v(x, y) = 0. \quad (6.2.31)$$

One can actually consider (6.2.31) to be

$$\frac{d}{dy} v(x(y), y) = 0 \quad (6.2.31')$$

since x is a function of y for the purposes considered here. If $x \neq y$, $E(x, y)$ is constant, so that if one differentiates (6.2.28) the following equation is obtained:

$$[f(x) - E(x, y)] \frac{\partial^2 v}{\partial x \, \partial y} - \delta \frac{\partial v}{\partial y} = 0. \quad (6.2.32)$$

Thus if $\partial v(y, y)/\partial y = 0$, $\partial v(x, y)/\partial y$ vanishes identically in x. Differentiating (6.2.29) and using the chain rule gives

$$\frac{\partial}{\partial x} v(y, y) + \frac{\partial v}{\partial y}(y, y) = \frac{1}{\delta} \frac{d}{dy} [n(y)f(y)]. \quad (6.2.33)$$

Using (6.2.30) gives

$$\frac{\partial}{\partial y} v(y, y) = \frac{1}{\delta} \frac{d}{dy} [n(y) f(y)] - n(y) = 0. \quad (6.2.34)$$

The optimal y thus satisfies the first-order condition

$$\frac{1}{\delta} \frac{d}{dy} [n(y) f(y)] = n(y). \quad (6.2.35)$$

The second-order condition is that $\partial^2 v(y, y)*\partial y^2 < 0$ (verify it!). Equation (6.2.35) is the same as the deterministic results in Chapter 3. The solution y^* of (6.2.35) is the optimal switching point in the deterministic case.

For the stochastic case assume that the solution of (6.2.22) has the following perturbation expansion valid for small ε:

$$v(x, y, \varepsilon) = \sum_{k=0} \varepsilon^k v_k(x, y). \quad (6.2.36)$$

Here the functions $v_k(x, y)$ are to be determined. After evaluating derivatives and substituting into (6.2.22), set the coefficients of powers of ε equal to 0. For the order 1 and order ε terms, this procedure gives

$$\begin{aligned} O(\varepsilon^0): & \quad g v_{0x} - \delta v_0 + nE = 0, \\ O(\varepsilon^1): & \quad D v_{0xx} + g v_{1x} - \delta v_1 = 0. \end{aligned} \quad (6.2.37)$$

In addition to (6.2.37), continuity of the solution requires that v_0, v_{0x}, v_1, and v_{1x} are continuous at $x = y$.

One can identify $v_0(x, y)$ with the previously computed deterministic solution as long as $v_0(y, y) = f(y) n(y)/\delta$. To do this, evaluate the equation for v_0 at $x > y$ and $x < y$, then subtract and take the limit as $x \to y$. This gives $v_{0x}(y, y) = n(y)$, and substituting back into the equation gives $v_0(y, y) = f(y) n(y)/\delta$. Thus, $v_0(x, y)$ is the deterministic solution.

The optimal y, y_ε^*, will satisfy a generalization of (6.2.31). This generalization is

$$\frac{\partial}{\partial y} v(x, y_\varepsilon^*, \varepsilon) = 0. \quad (6.2.38)$$

If $x \neq y$, differentiate (6.2.22) with respect to y to obtain

$$\varepsilon D v_{yxx} + g v_{yx} - \delta v_y = 0. \quad (6.2.39)$$

Equation (6.2.39) reduces to (6.2.32) when ε is set equal to 0. Arguing as before about the switching point shows that $\partial v(x, y)/\partial y \equiv 0$ in x if $\partial v(y, y)/\partial y = 0$. This shows that y_ε^* is independent of x. Thus, one can approximately obtain y_ε^* by solving the equation

$$v_{0y}(y^*, y_\varepsilon^*) + \varepsilon y_{1y}(y^*, y_\varepsilon^*) = 0. \quad (6.2.40)$$

6.2. HARVESTING A RANDOMLY FLUCTUATING POPULATION

Now assume that y_ε^*, the optimal switching point, is a perturbation of the deterministic switching point y^*. Then y_ε is modeled by

$$y_\varepsilon^* = y^* + dy^* \tag{6.2.41}$$

and dy^* must be found. Substituting (6.2.41) into (6.2.40) and expanding gives

$$\frac{\partial v_0}{\partial y}(y^*, y^*) + \frac{\partial^2 v_0}{\partial y^2}(y^*, y^*)\, dy^* + \varepsilon \frac{\partial v_1}{\partial y}(y^*, y^*) + O(\varepsilon^2) = 0. \tag{6.2.42}$$

The first term is identically zero, so that to leading order dy^* is given by

$$dy^* = -\frac{\varepsilon v_{1y}(y^*, y^*)}{v_{0yy}(y^*, y^*)}. \tag{6.2.43}$$

Equation (6.2.43) shows that once one knows $v_{1y}(y^*, y^*)$ and $v_{0yy}(y^*, y^*)$, the stochastic switching point is known, at least to order ε.

If $x \neq y$, differentiate the first equation in (6.2.37) with respect to x to obtain

$$gv_{0xx} + g_x v_{0x} - \delta v_{0x} + \frac{\partial}{\partial x}[n(x)E(x, y)] = 0. \tag{6.2.44}$$

Evaluate (6.2.44) for $x > y$ and $x < y$, let $x \to y$, and use $v_{0x}(y, y) = n(y)$ to obtain the two equations

$$g_\pm v_{0xx}^\pm + (\partial g_\pm/\partial x - \delta)n + n'E_\pm + n\, \partial E_\pm/\partial x = 0. \tag{6.2.45}$$

Here $\pm$ indicates the appropriate limit (from above or below).
Using the definition $g(x, y) = f(x) - E(x, y)$, (6.2.45) simplifies to

$$g_\pm v_{0xx}^\pm + (f' - \delta)n + E_\pm n' = 0, \tag{6.2.46}$$

and using (6.2.35) in (6.2.46) gives

$$g_\pm(v_{0xx}^\pm - n') + \delta v_{0y} = 0. \tag{6.2.47}$$

Now since $g_+ \neq g_-$ in general, v_{0xx} will be discontinuous at $x = y$ unless $v_{0y} = 0$, which occurs at $y = y^*$ (right where it is needed). Thus, at $x = y = y^*$, (6.2.47) shows that

$$v_{0xx}(y^*, y^*) = n'(y^*). \tag{6.2.48}$$

The next task is to find v_{1y}. To do this, take the limits $x \uparrow y$ and $x \downarrow y$ of the second equation in (6.2.37) to obtain

$$Dv_{0xx}^\pm + g_\pm v_{1x} - \delta v_1 = 0. \tag{6.2.49}$$

6. FLUCTUATING RENEWABLE RESOURCE STOCKS

These are two equations for v_{1x} and v_1 evaluated at $x = y$. Solving for v_{1x} gives

$$v_{1x} = \frac{-D(v_{0xx}^+ - v_{0xx}^-)}{g_+ - g_-}. \tag{6.2.50}$$

In light of (6.2.48), (6.2.50) implies

$$v_{1x}(y^*, y^*) = 0, \tag{6.2.51}$$

so that

$$\frac{d}{dy} v_1(y, y)\big|_{y^*} = v_{1y}(y^*, y^*). \tag{6.2.52}$$

Equation (6.2.52) shows that the total derivative of v_1 at (y^*, y^*) is simply the partial derivative with respect to y. To find the derivative v_{1y}, differentiate (6.2.49) with respect to y to obtain

$$\frac{d}{dy}(Dv_{0xx}^\pm) + g'_\pm v_{1x} + g_\pm \frac{d}{dy} v_{1x} - \delta \frac{d}{dy} v_1 = 0. \tag{6.2.53}$$

Eliminating the terms involving dv_{1x}/dy gives

$$\frac{1}{g_+} \frac{d}{dy}(Dv_{0xx}^+) - \frac{1}{g_-} \frac{d}{dy}(Dv_{0xx}^-) + \left(\frac{g'_+}{g_+} - \frac{g'_-}{g_-}\right) v_{1x} - \delta \left(\frac{1}{g_+} - \frac{1}{g_-}\right) \frac{dv_1}{dy} = 0. \tag{6.2.54}$$

Setting $y = y^*$, using $v_{1x}(y^*, y^*) = 0$, and solving gives

$$\frac{d}{dy} v_1\big|_{y^*} = \frac{\dfrac{1}{g_+}\dfrac{d}{dy}(Dv_{0xx}^+) - \dfrac{1}{g_-}\dfrac{d}{dy}(Dv_{0xx}^-)}{\delta\left(\dfrac{1}{g_+} - \dfrac{1}{g_-}\right)}\bigg|_{y^*}. \tag{6.2.55}$$

To find $dv_{0xx}^\pm/dy$, differentiate (6.2.47) with respect to y to obtain

$$g_\pm \frac{d}{dy}(v_{0xx}^\pm - n') + \left(\frac{d}{dy} g_\pm\right)(v_{0xx}^\pm - n') + \delta \frac{d}{dy} v_{0y} = 0. \tag{6.2.56}$$

At $y = y^*$ the middle term vanishes, since $v_{0xx}(y^*, y^*) = n'(y^*)$; thus

$$\frac{d}{dy} v_{0xx}^\pm\big|_{y^*} = \left[n'' - \frac{\delta}{g_\pm} \frac{d}{dy} v_{0y}\right]\bigg|_{y^*}. \tag{6.2.57}$$

6.2. HARVESTING A RANDOMLY FLUCTUATING POPULATION

Putting all of this together gives

$$dy^* = \varepsilon D \left[\frac{1}{g_+} + \frac{1}{g_-} \right]\bigg|_{y^*} + \varepsilon \left[\frac{(Dn')'}{-\delta \, dv_{0y}/dy} \right]\bigg|_{y^*} + O(\varepsilon^2). \quad (6.2.58)$$

The first term in (6.2.58) is due to the constraints on effort. It is such that its effect is to move y^* so that $f(y^*)$ is not too close to either E_{max} or E_{min}. The second term in (6.2.58) is due to changes in the net revenue caused by fluctuations around the switching point.

The result derived here is only an approximation, since the basic ansatz (6.2.36) cannot be uniformly valid over the entire range of stocks and is valid only for small ε. Ludwig (1979a) also considers the much harder problem of three switching points and a number of other extensions of the basic model presented here. Ludwig and Varah (1979) develop numerical methods for the application of this theory.

In a separate paper Ludwig (1979b) discusses the same kind of problem when there is a cost involved with switching effort; this kind of assumption reduces some of the bang-bang nature of the control (see Exercise 6.5.1).

Another approach for harvesting a randomly fluctuating population is described by Ludwig (1980). The starting point there is the population dynamics equation

$$dX = [f(X) - X - qhX] \, dt + \sqrt{2\varepsilon X^2} \, dW. \quad (6.2.59)$$

Here h is a measure of effort, q is the catchability coefficient, and $f(X) - X$ represents the stock dynamics.

Exercise 6.2.11

How is the stock dynamics equation (6.2.59) related to the other versions, (6.2.1) and (6.2.17)?

Three possible forms for $f(X)$ are

$$f(X) = 3X/(1 + 2X) \quad \text{(Beverton and Holt, 1957),}$$
$$f(X) = X(1 - X) + X \quad \text{(logistic),} \quad (6.2.60)$$
$$f(X) = \tfrac{2}{3}X(1 - X^2) + X \quad \text{(Pella and Tomlinson, 1969).}$$

Exercise 6.2.12

First, analyze these three forms for $f(X) - X$. Show that they all have a carrying capacity of 1 and that they match at $f(0)$ and $f(0.5)$. What is the

6. FLUCTUATING RENEWABLE RESOURCE STOCKS

maximum sustainable yield (MSY) for each model? Second, analyze the behavior of

$$dX/dt = f(X) - X.$$

As control objectives Ludwig considers the following three functionals:

(1) $J_1(x; s) = E\left\{\int_s^\infty e^{-\delta t} qh(t)X(t)\, dt \,|\, X(s) = x\right\},$

(2) $J_2(x, s) = \dfrac{[E\{\int_s^\infty e^{-\delta t} qh(t)X(t)\, dt\,|\, X(s) = x\}^2 - J_1(x, s)^2]^{1/2}}{J_1(x)},$ (6.2.61)

(3) $\quad J_3(x) = E\{\min t: X(t) = 0.1,\ X(s) > 0.1,\ s > t \,|\, X(0) = x\}.$

The first functional is the expected harvest at constant price p set equal to 1, the second is the coefficient of variation of that harvest, and the third is the mean time to reach 10% of the carrying capacity.

It is straightforward to show that if $J_1(x, s) = v(x)e^{-\delta s}$, then $v(x)$ satisfies

$$\varepsilon x^2 \frac{d^2 v}{dx^2} + [f(x) - x - qhx]\frac{dv}{dx} - \delta v + qhx = 0. \qquad (6.2.62)$$

Exercise 6.2.13

Verify (6.2.62).

In a similar way, if $J_2(x, s) = w(x)e^{-2\delta s}$, then $w(x)$ satisfies

$$\varepsilon x^2 \frac{d^2 w}{dx^2} + [f(x) - x - qhx]\frac{dw}{dx} - 2\delta w + 2qxhv(x, h) = 0. \qquad (6.2.63)$$

Finally, it can be shown that $J_3(x)$ satisfies

$$\varepsilon x^2 \frac{d^2 J_3}{dx^2} + [f(x) - x - qhx]\frac{dJ_3}{dx} = -1. \qquad (6.2.64)$$

Exercise 6.2.14

Verify (6.2.63) and (6.2.64).

The following four harvesting strategies are considered by Ludwig (1980).

(1) *Constant effort.* Here h is held constant throughout the harvesting period and chosen to maximize the expected discounted yield.

6.2. HARVESTING A RANDOMLY FLUCTUATING POPULATION

(2) *Optimal effort.* Here h is chosen according to a DP algorithim. Thus, in light of (6.2.62), for example, for maximizing the first functional the optimal effort is given by

$$h = \begin{cases} h_{\max} & \text{if } 1 > dv/dx, \\ h_{\min} & \text{if } 1 < dv/dx, \end{cases} \quad (6.2.65)$$

where $h_{\max}$ and $h_{\min}$ are the maximum and minimum possible values for the harvest rates. Ludwig uses $h_{\min} = 0$ and $h_{\max} = 2$.

(3) *Strategy of the International Whaling Commission (IWC).* In this case the maximum allowable catch is 90% of the maximum sustainable yield (MSY) if the population exceeds the MSY. The quota is decreased by 10% for every 1% that the population is below the MSY.

(4) *Constrained Optimal.* Here

$$h = \begin{cases} 2, & x > \hat{x}, \\ 0, & x < \hat{x}, \end{cases}$$

where $\hat{x}$ is chosen so that the time to reach 10% of the carrying capacity is at least as long as the time using the IWC strategy.

Exercise 6.2.15

Under what conditions would the constrained optimal strategy (4) be optimal?

Exercise 6.2.16

Before proceeding try to rank the four different strategies for each of the objective functionals.

Ludwig further considers the case in which there is a cost of switching effort given by

$$\beta \int_0^\infty e^{-\delta t}(dh/dt)^2 \, dt, \quad (6.2.66)$$

and also derives results for this case.

Ludwig's numerical results are shown in Table 6.2.1 for $x = 0.5$, $\delta = 0.1$, and $s = 0$.

Exercise 6.2.17

How can the results in Table 6.2.1 be compared to determine the "best" control policy. What would best mean in this case?

Table 6.2.1

Results of the Dynamic Programming Calculations[a]

Noise intensity ε	Dynamics strategy–criterion	Beverton–Holt			Logistic			Pella–Tominson		
		J_1	J_2	J_3	J_1	J_2	J_3	J_1	J_2	J_3
0.2	Constant effort	2.5	0.19	11	1.8	0.29	11	1.5	0.37	12
	Optimal	2.8	0.17	140	2.4	0.27	72	2.2	0.33	51
	IWC	1.8	0.13	350	1.3	0.22	126	1.2	0.27	76
	Constrained optimal	2.6	0.23	390	2.2	0.33	129	2.0	0.39	76
0.4	Constant effort	2.1	0.30	3.9	1.30	0.47	4.6	0.84	0.63	4.8
	Optimal	2.6	0.26	8.0	2.00	0.41	7.3	1.70	0.54	6.9
	IWC	1.4	0.21	15.0	0.93	0.35	11.0	0.73	0.47	9.0
	Constrained optimal	2.4	0.36	15.0	1.80	0.54	11.0	1.50	0.64	8.9

[a] From Ludwig (1980).

Bibliographic Notes

The use of a single stock, rather than an age-structured stochastic model, throughout this section shows the difficulty of these problems. Turelli (1977) discusses various issues involved with the modeling of stochastic effects. Asymptotic methods, such as the ones used by Ludwig (1979a), are discussed in Lin and Segel (1974) and Nayfeh (1973). A good introduction to the use of such methods in population dynamics is found in Ludwig (1975). The book of Beverton and Holt (1957) provides a wealth of material related to fisheries, as does Ricker (1975). Bang-bang control policies are discussed in Clark (1976) and Kamien and Schwartz (1981). A good reference on partial differential equations is the book by John (1978). The strategy of the International Whaling Commission is discussed in May et al. (1978). The last set of calculations in this section, in which different controls are ranked differently for various objective functions, highlights the fact that in most resource problems there are conflicting objectives. The books by Bell (1977) and Janis and Mann (1977) provide an introduction to the problems of conflicting decisions.

6.3 DEALING WITH PARAMETER UNCERTAINTY IN MANAGED POPULATIONS

The results of Sections 6.1 and 6.2 can be faulted because all the parameters of the system are assumed to be known with certainty. In most, if not all situations, this will not be true. As the population is managed, information about system parameters is gained and a management strategy should include the informational as well as the harvest component. Walters and

6.3. PARAMETER UNCERTAINTY IN MANAGED POPULATIONS

Ludwig (1980) and Ludwig and Walters (1981, 1982) have considered this problem for salmonid populations. Their work is summarized in this section.

In salmonid populations the key problem is to assess the relationship between spawners in generation j (s_j) and recruits in the next generation (r_{j+1}). For simplicity (although the results are robust to the form of the stock–recruitment relationship), the Ricker formula is used to model the stock–recruitment relation. The deterministic version is

$$r_{j+1} = s_j \exp\{a + bs_j\}. \tag{6.3.1}$$

For a biologically reasonable model, it must be that $a > 0$, $b < 0$.

Exercise 6.3.1

The basic model (6.3.1) is used throughout this section. It must be thoroughly understood.

Consequently,

(i) discuss the interpretations of a and b,
(ii) compute the equilibrium population,
(iii) discuss the effect of a harvest that removes a fraction h of the population.

For the stochastic version with parameter uncertainty, replace (6.3.1) by

$$r_j = s_{j-1} \exp\{a + bs_{j-1} + u_j\}, \tag{6.3.2}$$

where a, b are parameters to be estimated and $u_j \sim N(0, \sigma_u^2)$ are noise terms. Equation (6.3.2) characterizes the true numbers of spawners and recruits. The observed values are assumed to be a stochastic perturbation of the true values. In particular, the observed numbers of recruits and spawners are assumed to follow

$$R_j = r_j e^{v_j}, \qquad S_j = s_j e^{v_j}. \tag{6.3.3}$$

Here the $v_j \sim N(0, \sigma_v^2)$ are measurement noise terms. The question is then, given a data set $\{R_1, S_1, R_2, S_2, \ldots, R_N, S_N\}$, estimate a, b, σ_u^2, and $\{v_j\}$. Assume [otherwise the problem is insoluble; see Kendall and Stewart (1973) and Exercise 6.3.5] that

$$\sigma_v^2 = \lambda \sigma_u^2. \tag{6.3.4}$$

The value of λ must be specified in advance of all numerical calculations. Numerical results reported in Ludwig and Walters (1981) and Ludwig and Hilborn (1983) appear to be relatively robust over a wide range of values for λ. With this assumption a, b, λ, and $\{v_j\}$ are to be determined. Combining (6.3.2) and (6.3.3) gives

$$R_j \exp(-v_j) = S_{j-1} \exp(-v_{j-1}) \exp[a + bS_{j-1} \exp(-v_{j-1}) + u_j], \quad (6.3.5)$$

so that

$$u_j = -a - bS_{j-1}e^{-v_{j-1}} + v_{j-1} - v_j + \ln(R_j/S_{j-1}). \quad (6.3.6)$$

If one sets $Y = \ln(R_j/S_{j-1})$, $X_{j1} = 1$, and $X_{j2} = S_{j-1}$, then (6.3.6) can be put into the form of a linear regression:

$$Y_j = \sum_{p=1}^{2} X_{jp} a_p + u_j - v_{j-1} + v_j. \quad (6.3.7)$$

Exercise 6.3.2

What are the a_p? Apply linear regression theory to estimate the parameters in (6.3.7).

Exercise 6.3.3

Are there any sources of bias in this linear regression formulation?

The advantage of (6.3.7) is that the nonlinear dynamics have been converted into the form of a linear regression problem (Draper and Smith, 1981). Walters and Hilborn (1976) and Smith and Walters (1981) made considerable use of this formulation.

A somewhat different formulation works directly with the information contained in (6.3.6) instead of the regression form (6.3.7). To do this, first introduce subjective weights for each of the observations.

Let w_j be a subjective weight for the jth observation. In light of the assumptions on u_j and v_j, the likelihood of the observations (6.3.6) is given by (recall the normality assumption)

$$L(a, b, v|\lambda) = \exp\left\{-\frac{1}{\sigma_u^2}\left(\frac{1}{2}\sum_{j=1}^{N} w_j u_j^2 + \frac{1}{2\lambda}\sum_{j=0}^{N} w_{j+1} v_j^2\right)\right.$$
$$\left. - \sum_{j=1}^{N} w_j \ln \sigma_u - \sum_{j=0}^{N} w_{j+1} \ln \sigma_v\right\}. \quad (6.3.8)$$

Exercise 6.3.4

Derive (6.3.8) from the above assumptions and Eqs. (6.3.1) and (6.3.3).

6.3. PARAMETER UNCERTAINTY IN MANAGED POPULATIONS

Exercise 6.3.5

Suppose that the assumption (6.3.4) were not made. Reformulate the likelihood using both σ_u^2 and σ_γ^2 as parameters. What difficulties occur if one tries to maximize this new likelihood?

For ease of notation in what follows, set

$$a = (a_1, a_2) = (a, b), \qquad v = (v_1, v_2, \ldots, v_N), \tag{6.3.9}$$

$$M(a, v) = \frac{1}{2}\sum_{j=1}^{N} w_j u_j^2 + \frac{1}{2\lambda}\sum_{j=0}^{N} w_{j+1} v_j^2, \tag{6.3.10}$$

so that (6.3.6) becomes

$$L(a, v | \lambda) = \exp\left\{ -\frac{M(a,v)}{\sigma_u^2} - \sum_{j=1}^{N} w_j \ln \sigma_u - \sum_{j=0}^{N} w_{j+1} \ln \sigma_v \right\}. \tag{6.3.11}$$

Equation (6.3.11) is the likelihood function for the data with unknown parameters a_1, a_2, and $v_1, \ldots, v_N$. The problem now is to determine the values of the parameters. A maximum likelihood approach can be used. Maximizing (6.3.11) with respect to (a, v) and using the notation introduced in (6.3.7) gives the first-order conditions at the optimal estimates $(\hat{a}, \hat{v})$:

$$\left.\frac{\partial M}{\partial a_p}\right|_{\hat{a},\hat{v}} = \sum_{j=1}^{N} w_j \hat{u}_j \hat{X}_{jp} = 0,$$

$$\left.\frac{\partial M}{\partial v_k}\right|_{\hat{a},\hat{v}} = \sum_{j=1}^{N} w_j \hat{u}_j \frac{\partial \hat{u}_i}{\partial v_k} + \frac{1}{\lambda} w_{k+1} \hat{v}_k = 0. \tag{6.3.12}$$

Exercise 6.3.6

Derive (6.3.12). How would they compare to results obtained by regression methods?

Combining these two equations gives

$$\left.\frac{\partial M}{\partial v_k}\right|_{\hat{a},\hat{v}} = w_{k+1}\left\{\hat{u}_{k+1}\frac{\partial \hat{u}_{k+1}}{\partial v_k} + \frac{1}{\lambda} v_k\right\} + w_k \hat{u}_k \frac{\partial \hat{u}_k}{\partial v_k} = 0. \tag{6.3.13}$$

From (6.3.6) it follows that

$$\partial u_{k+1}/\partial v_k = 1 + a_2 s_k, \qquad \partial u_k/\partial v_k = 1. \tag{6.3.14}$$

Equations (6.3.12)–(6.3.14) provide a set of nonlinear equations for estimating the parameters. Ludwig and Walters (1981) solve them by Newton's method (see Chapter 8 for a discussion of Newton's method). All the details of this solution are found in their paper.

The solution of the system (6.3.12)–(6.3.14) does not provide the whole story, though. The parameters obtained by maximizing the likelihood depend upon the data. Consequently, one must consider the question of the accuracy of the parameter estimates.

To obtain a measure of the accuracy of the parameter estimates, think of v_j as a function of the parameter a. Now if the first-order condition

$$\frac{\partial M}{\partial v_k}(a, v(a)) = 0 \qquad (6.3.15)$$

is differentiated with respect to a_p, one obtains

$$\frac{\partial^2 M}{\partial a_p \, \partial v_k} + \sum_{l=0}^{N} \frac{\partial^2 M}{\partial v_k \, \partial v_l} \frac{\partial v_l}{\partial a_p} = 0. \qquad (6.3.16)$$

The quantities $\partial v_l / \partial a_p \equiv -h_{pl}$ arise in the solution of (6.3.12) and (6.3.13) by Newton's method. Now, in addition, the second-order a derivatives are

$$\frac{\partial^2 M(a, v(a))}{\partial a_p \, \partial a_q} = \frac{\partial^2 M}{\partial a_p \, \partial a_q} + \sum_{k=0}^{N} \frac{\partial^2 M}{\partial a_p \, \partial v_k} \frac{\partial v_k}{\partial a_p} \equiv M'_{pq}. \qquad (6.3.17)$$

An approximation for $M(a, v(a))$ around the MLE estimates (using a Taylor expansion) is

$$M(a, v(a)) \simeq M(\hat{a}, \hat{v}(\hat{a})) + \frac{1}{2} \sum_{p,q} M'_{pq}(a_p - \hat{a}_p)(a_q - \hat{a}_q). \qquad (6.3.18)$$

Equation (6.3.18) provides a measure of the accuracy of the parameter estimates. At this level of approximation, the likelihood is a multivariate normal density. In particular, if one uses (6.3.18) in (6.3.11), the resulting likelihood can be viewed as the likelihood of the parameters, conditioned on the data. As such it summarizes the information about the parameters that is contained in the data.

Once these estimates $(\hat{a}, \hat{v})$ are known, the unbiased estimate of σ_u^2 is obtained from the residual sum of squares. It is

$$\hat{\sigma}_u^2 = \frac{1}{N-2} \left\{ \sum_{i=1}^{N} \hat{u}_i^2 + \frac{1}{\lambda} \sum_{i=0}^{N} \hat{v}_j^2 \right\}. \qquad (6.3.19)$$

Equations (6.3.17)–(6.3.19) provide a way of dealing with the accuracy of the parameter estimates.

Exercise 6.3.7

Show that if σ_u were determined by maximizing the likelihood, then the estimate would be biased.

6.3. PARAMETER UNCERTAINTY IN MANAGED POPULATIONS

What about management in the presence of parameter uncertainty? Ludwig and Walters (1981) consider a simple escapement policy for harvest. The harvest h_j is given by

$$h_j = \max\{r_j - q, 0\}, \tag{6.3.20}$$

where q is the targeted escapement level. The equilibrium harvest level is then (with R denoting the equilibrium population of recruits)

$$h = R - q. \tag{6.3.21}$$

If parameter uncertainty is ignored, then the expected harvest is

$$\hat{H}(q) = E_u\{q \exp(\hat{a} + \hat{b}q + u)\} - q = q[\exp(\hat{a} + \hat{b}q + \tfrac{1}{2}\sigma_u^2 - 1]. \tag{6.3.22}$$

Exercise 6.3.8

Derive equation (6.3.22). Compute a formula for the optimal escapement level q.

One way to include parameter uncertainty is to assume that the real parameters a, b are normally distributed with mean $\hat{a}, \hat{b}$ and covariance

$$\sigma_u^2 \begin{bmatrix} M'_{aa} & M'_{ab} \\ M'_{ab} & M'_{bb} \end{bmatrix}^{-1} \tag{6.3.23}$$

(this is the inverse of the Fisher information matrix). In this case, instead of (6.3.22), the equilibrium harvest is

$$\hat{H}(q) = q\{\exp[\hat{a} + \hat{b}q + \tfrac{1}{2}(\sigma_u^2 + \sigma_z^2(q))] - 1\}, \tag{6.3.24}$$

where $z = (a - \hat{a}) + (b - \hat{b})S$ and $\sigma_z^2(q)$ is the variance of z at harvest rate q.

Exercise 6.3.9.

Compare the two expected harvest rates. What can be said about the nature of the solutions?

Ludwig and Walters (1981) use these results to analyze salmon data and also show how the results can be used with other stock–recruitment models.

Exercise 6.3.10

An alternative plan is to consider a harvest that takes a proportion of the spawners. Let $1 - q$ denote this proportion. Then there remains a proportion q of the spawners which give rise to recruits. In the absence of noise or uncertainty, the number of recruits under this harvest policy is

$$R = qS[\exp(a + bqS)]. \tag{6.3.25}$$

For this model compute the equilibrium harvest under no parameter uncertainty [analogous to (6.3.21)] and with parameter uncertainty [analogous to (6.3.24)].

Bibliographic Notes

When parameters are unknown, control action provides probing about parameter values as well as returns from the system. This idea is developed by Bar Shalom and Tse (1976) in general and by Ludwig and Hilborn (1983), Silvert (1977), Smith and Walters (1981), Walters (1981), and Walters and Hilborn (1976) for fishing systems. Alternatives to the Ricker model are discussed in Cushing (1981). Draper and Smith (1981) provide a good introduction to linear regression. The situation in which the independent variable is measured with noise is called "errors in variables." It is highly likely that such errors are present in data from almost every fishery. Maximum likelihood estimation is discussed by Sorenson (1980). Newton's method is discussed in almost every text on numerical analysis and Section 8.1 of this book. Some references are Blum (1972), Burden, Faires, and Reynolds (1978), and Stoer and Bulirsch (1980). The use of residual sum of squares is discussed by Kendall and Stuart (1973) as well as Draper and Smith (1981).

6.4 PRICE DYNAMICS IN RENEWABLE RESOURCE MARKETS

As in the case of exhaustible resource markets, the price dynamics of the renewable resource market can be studied by the techniques developed in this book. The deterministic analogs were studied in Chapter 3. The approach used here is similar to Pindyck (1984).

The starting point is the stochastic dynamic equation for the stock biomass $X(t)$. This equation is assumed to be

$$dX = [f(X) - q]\, dt + \sigma(X)\, dW. \tag{6.4.1}$$

Here $f(x)$ is the deterministic rate of growth in the absence of harvest, q is the rate of harvest, $\sigma(x)$ measures the intensity of the fluctuations, and $W(t)$ is standard Brownian motion with $E\{dW\} = 0$, $E\{(dW)^2\} = dt$. If p is the price obtained per unit biomass and $c(x)$ the per unit harvest cost when the biomass level is x, then the objective functional of interest is

$$J(x, s) = \max_{q} E\left\{ \int_{s}^{\infty} [p - c(x)]q e^{-\delta t}\, dt \,\Big|\, X(0) = x \right\}. \tag{6.4.2}$$

6.4. PRICE DYNAMICS IN RENEWABLE RESOURCE MARKETS

Exercise 6.4.1

After deriving the DPE for $J(x, s)$, set $J(x, s) = e^{-\delta s}v(x)$ to obtain for $v(x)$ the time-independent equation

$$0 = -\delta v + \max_q \{[p - c(x)]q + [f(x) - q]v_x + \tfrac{1}{2}\sigma^2(x)v_{xx}\}. \quad (6.4.3)$$

What would the DPE be for an arbitrary function $u(p, c, x)$ instead of profit?

The optimal harvest rate, as before, is a bang-bang one, given by

$$q^* = \begin{cases} q_{\max} & \text{if } p - c(x) > v_x, \\ q_{\min} & \text{if } p - c(x) < v_x. \end{cases} \quad (6.4.4)$$

Here $q_{\max}$ and $q_{\min}$ are the maximum and minimum allowable harvest rates. The singular path is characterized by the condition

$$p - c(x) = v_x. \quad (6.4.5)$$

Exercise 6.4.2

Explain why v_x measures the marginal value of a unit biomass at stock level x. What is the appropriate interpretation of the integrated form of (6.4.5)?

In anticipation of the study of price dynamics, note that (6.4.5) also implies the differential relationship

$$E\{d(p - c(x))\} = E\{d(v_x)\}. \quad (6.4.6)$$

Understanding q to be q^* and differentiating (6.4.3) with respect to x gives the partial differential equation

$$0 = -\delta v_x - c'(x)q + f'(x)v_x + f(x)v_{xx} + \tfrac{1}{2}\sigma^2(x)v_{xxx} + \sigma(x)\sigma'(x)v_{xx}. \quad (6.4.7)$$

Equation (6.4.7) can be rewritten as

$$0 = \lim_{dt \to 0} \left[-\delta v_x - c'(x)q + \sigma(x)\sigma'(x)v_{xx} + f'(x)v_x + \frac{1}{dt}E\{d(v_x)\} \right]. \quad (6.4.8)$$

Solving (6.4.8) for $E\{d(v_x)\}$ and using it in (6.4.6) gives an equation for the price dynamics. This equation is

$$\lim_{dt \to 0} \frac{1}{dt} \frac{E\{d(p - c(x))\}}{p - c(x)} = \delta - f'(x) + \frac{c'(x)q}{p - c(x)} - \frac{\sigma(x)\sigma'(x)v_{xx}}{p - c(x)}. \quad (6.4.9)$$

Comparing this with the deterministic result in Chapter 3, one sees an additional term, due to stock fluctuations, is present. If $\sigma'(x) \equiv 0$, then the right-hand side of (6.4.9) is the same as the deterministic result derived in Chapter 3.

Exercise 6.4.3

If $\sigma'(x) = 0$, does (6.4.9) imply that the price dynamics are essentially deterministic? If $\sigma'(x) \neq 0$, what can be said about (6.4.9)?

The last term on the right-hand side of (6.4.9) can be interpreted as follows. In light of (6.4.5), the last term in (6.4.9) is proportional to

$$-v_{xx}/[p - c(x)] = -v_{xx}/v_x \equiv A(x), \qquad (6.4.10)$$

which is a measure of risk aversion. In the risk-averse case [for reasonable $c(x)$], $A(x) > 0$, so that the last term in (6.4.9) is positive. It can be interpreted as a risk premium.

The cost and price dynamics can be analyzed separately. For example, the cost dynamics evolve according to

$$\lim_{dt \to 0} \frac{1}{dt} \mathrm{E}\{dc(X)\} = \lim_{dt \to 0} \frac{1}{dt} \mathrm{E}\{c'(x)\,dX + \tfrac{1}{2}c''(x)\,dX^2 + o(dt)\} \quad (6.4.11)$$

$$= c'(x)[f(x) - q] + \tfrac{1}{2}c''(x)\sigma(x)^2, \qquad (6.4.12)$$

so that average cost does not match the deterministic cost, even if $\sigma'(x) = 0$.

Exercise 6.4.4

Interpret the second term in (6.4.12).

Using (6.4.12) in (6.4.9) gives the following for the price dynamics:

$$\lim_{dt \to 0} \frac{1}{dt} \mathrm{E}\{dp\} = c'(x)f(x) + \tfrac{1}{2}c''(x)\sigma(x)^2 + \delta[p - c(x)]$$

$$- f'(x)[p - c(x)] + \sigma(x)\sigma'(x)A(x)[p - c(x)]. \qquad (6.4.13)$$

Exercise 6.4.5

Interpret (6.4.13) and compare the result to the deterministic analog.

It is not clear what a price–stock equilibrium would mean in this context, since the stochastic fluctuations would always move the system out of equilibrium. An equilibrium point of $\mathrm{E}\{dX\}$ and $\mathrm{E}\{dp\}$ can, of course, be found.

6.4. PRICE DYNAMICS IN RENEWABLE RESOURCE MARKETS

Exercise 6.4.6

Of more interest than the equilibrium point would be the two-dimensional equilibrium probability density for stock and price. How can it be calculated?

In order to calculate the equilibrium probability density $\rho(X, p)$, one needs to solve the following forward equation (see Chapter 1) for $\rho(x, p)$:

$$\frac{1}{2}\left[\frac{\partial^2}{\partial x^2}(\sigma^2(x)\rho) + 2\frac{\partial^2}{\partial x\, \partial p}(\sigma^2_{px}\rho) + \frac{\partial^2}{\partial p^2}(\sigma^2_{pp}\rho)\right]$$
$$-\frac{\partial}{\partial x}\{[f(x) - q]\rho\} - \frac{\partial}{\partial p}(\mu_p \rho) = 0, \qquad (6.4.14)$$

where

$$\mu_p = \lim_{dt\to 0} \frac{1}{dt} E\{dp\},$$

$$\sigma^2_{pp} = \lim_{dt\to 0} \frac{1}{dt} E\{(dp)^2\},$$

$$\sigma^2_{px} = \lim_{dt\to 0} \frac{1}{dt} E\{dp\, dx\}.$$

When does the stationary density $\rho(x, p)$ exist? How would one find it? If possible, solve (6.4.14).

Exercise 6.4.7

Pindyck (1984) studies the particular case

$$\sigma(x) = \sigma_0 x, \qquad q(p) = b/p^n, \qquad c(x) = c/x^\gamma$$

(the second equation characterizes the demand function) and picks $f(x) = rx(1 - x)$, $f(x) = rx \log x$, or $f(x) = r\sqrt{x} - rx$ (here the carrying capacity is normalized to 1). In each case it is possible to find the value function and the steady-state probability distribution of the population. This steady-state distribution, $\rho(x)$, satisfies

$$\frac{1}{2}\frac{\partial^2}{\partial x^2}(\sigma^2(x)\rho) - \frac{\partial}{\partial x}([f(x) - q]p) = 0. \qquad (6.4.15)$$

Obtain the value function, the steady-state probability distribution $\rho(x)$, and the steady-state average extraction rate [obtained from an equation analogous to (6.4.15) but using the moments of dq] for each choice of $f(x)$.

Discuss the effects of noise and uncertainty on the extraction rate and on the value function. (*Hint*: For $f(x) = rx(1 - x)$, choose $\eta = \frac{1}{2}$, $\gamma = 2$; for $f(x) = rx \ln x$, choose $\eta = \gamma = 1$; for $f(x) = r\sqrt{x} - rx$, choose $\eta = 2$, $\gamma = \frac{1}{2}$.)

Bibliographic Notes

The method used in this section is essentially the method of Arkin, Kolemaev, and Sirjaev (1966) that was discussed in Chapter 2. This method allows one to study the solution of a DPE without actually solving it. Questions of stability in stochastic systems, although very interesting, are usually quite difficult and technical. Readable introductions are provided by Arnold (1974) and Schuss (1980). Pindyck (1984) provides further interpretations and results. Ludwig (1975) discusses methods for solving equations similar to (6.4.14).

6.5 OPTIMAL TRADE-OFF BETWEEN ECONOMIC RETURN AND THE RISK OF UNDESIRABLE EVENTS

Often the optimal harvest rate obtained using the stochastic control techniques is a bang-bang control. Such controls have social costs since they are effectively "boom or bust" and require extensive capitalization that is often not utilized. This makes such solutions unattractive. Consequently, it is worthwhile to try to develop a theory in which control fluctuations are smoothed out. Ludwig (1979b) has done this for the stock equation analyzed in Section 6.2,

$$dX = \{f(x) - h\}\,dt + \sigma(x)\,dW, \qquad (6.5.1)$$

by using the objective functional

$$J(x, t) = \max_{h} \mathrm{E}\left\{\int_{t}^{\infty}\left[[p - c(x)]h - \beta\left(\frac{dh}{dt}\right)^{2}\right]e^{-\delta s}\,ds\,\middle|\,X(t) = x\right\}. \qquad (6.5.2)$$

Here $f(x)$ is the deterministic growth rate, h the harvest rate, $\sigma(x)$ the intensity of fluctuations, $W(t)$ a standard Brownian motion, δ the discount rate, and β a parameter measuring the cost of changing effort.

Exercise 6.5.1

If $\beta \neq 0$, why is it clear that a bang-bang control can never be optimal?

Ludwig (1979b) deals with (6.5.2) essentially by letting $h(t)$ be a new state variable and $w = dh/dt$ be the control variable.

6.5. TRADE-OFF BETWEEN ECONOMIC RETURN AND RISK

Exercise 6.5.2

What is the DPE for such a problem in (x, w) space? Why is it much harder to analyze than the DPEs studied previously?

Mendelssohn (1979, 1980) noted that bang-bang controls lead to widely fluctuating harvest rates, which may also be undesirable. His method for smoothing out fluctuations is different from Ludwig's and proceeds as follows.

In discrete time assume that x_t is the population in year t before the harvest, z_t is the harvest, and $y_t = x_t - z_t$. The dynamics of the stock are assumed to be given by

$$x_{t+1} = f(y_t, v_t), \tag{6.5.3}$$

where v_t are independent, identically distributed noise variables and $f(\cdot)$ is the growth function. With this notation the optimization problems that have been considered up to this point take the form

$$\text{maximize} \quad E\left\{\sum_{t=1}^{\infty} \alpha^{t-1} g(x_t, y_t)\right\}, \tag{6.5.4}$$

where $g(x_t, y_t)$ is the value function for a harvest of y_t at stock level x_t and α the discount rate. The strategy that maximizes (6.5.4) may lead to harvests that fluctuate greatly. To smooth out such fluctuations, one can assess a cost for fluctuations in the harvest. One way to do this is to replace (6.5.4) by a functional that explicitly excludes the cost of such fluctuations. One choice is

$$\text{maximize} \quad E\left\{\sum_{t=1}^{\infty} \alpha^{t-1} g(x_t, y_t) - \lambda|x_t - y_t - z_{t-1}|\right\}. \tag{6.5.5}$$

Here λ is a parameter that measures the cost of fluctuations in the harvest. As λ increases the optimal harvest policies will tend towards constant z_t. For the objective functional (6.5.5), Mendelssohn shows that the set of Pareto[†] optimal policies is the set of harvest policies such that if $g(x, y)$ increases, so does $\lambda|x - y - z|$ and if $g(x, y)$ decreases, so does $\lambda|x - y - z|$. Mendelssohn's approach is to consider two value functions by separating $g(x_t, y_t)$ and $-\lambda|x_t - y_t - z_{t-1}|$. In this way one can compute Pareto optimal policies in a highly efficient manner. Mendelssohn (1980) discusses numerical methods for efficient maximization and provides an example from the Alaska salmon fishery.

[†] A Pareto optimal policy is defined in the following way. Suppose that there are $n \geq 2$ value functions $v_i(x, u)$, where $i = 1, \ldots, n$. A control u^* is Pareto optimal if $v_i(x, u^*) \geq v_i(x, u)$ for all other u and all i and $v_i(x, u^*) > v_i(x, u)$ for at least one i.

A somewhat different theory, but in the same spirit, is the following. Let $X(t)$ be the state variable and u the control. Assume that the state dynamics are governed by

$$dX = b(X, u)\, dt + \sqrt{a(X, u)}\, dW. \tag{6.5.6}$$

Suppose that $X(0)$ is in the interval (l_1, l_2) and that the goal of the management strategy is to keep the system in (l_1, l_2) for as long as possible. Such a management strategy implicitly smooths out control and harvest fluctuations if the interval is small enough.

Let $T(x)$ be the exit time defined by

$$T(x) = \min_{t}\{t : X(t) \notin (l_1, l_2) | X(0) = x\}. \tag{6.5.7}$$

Since $T(x)$ is a random variable, one must deal with its expection:

$$\bar{T}(x) \equiv \mathrm{E}\{T(X)\} = \mathrm{E}\left\{\int_0^T dt \,|\, X(0) = x\right\}. \tag{6.5.8}$$

What is the differential equation for $\bar{T}(x)$? First treat a as given exogenously. Comparing $\bar{T}(x)$ and $\bar{T}(x + dx)$ gives

$$\bar{T}(x) = \mathrm{E}_{dX}\{\bar{T}(x + dX)\} + dt. \tag{6.5.9}$$

Taylor expanding and taking the moments in (6.5.9) gives

$$\bar{T}(x) = \bar{T}(x) + [b(x, u)\bar{T}'(x) + \tfrac{1}{2}a(x, u)\bar{T}''(x) + 1]\, dt + o(dt). \tag{6.5.10}$$

The equation for $\bar{T}(x)$ is then

$$-1 = b(x, u)\bar{T}'(x) + \tfrac{1}{2}a(x, u)\bar{T}''(x). \tag{6.5.11}$$

The problem of real interest is the optimal control problem.

Exercise 6.5.3

Show that the DPE for the objective function $\max_u \bar{T}(x)$ is

$$-1 = \max_{u}[b(x, u)\bar{T}'(x) + \tfrac{1}{2}a(x, u)\bar{T}''(x)]. \tag{6.5.12}$$

The general solution of (6.5.12), however, may still be a bang-bang control with large state fluctuations. To try to smooth them out, one can use the second representation of $\bar{T}(x)$ in (6.5.8) and introduce costs associated with stock and control fluctuations. In particular, assume that a desired stock level μ and control level $\bar{u}$ are given exogenously. Consider the functional

$$J(x, \mu, \bar{u}) = \max_{u} \mathrm{E}\left\{\int_0^T 1 - \alpha_1[X(t) - \mu]^2 - \alpha_2(u(t) - \bar{u})^2 \, dt \,|\, X(0) = x\right\}. \tag{6.5.13}$$

6.5. TRADE-OFF BETWEEN ECONOMIC RETURN AND RISK

Here α_1 and α_2 are the costs of stock and control fluctuations, respectively, and, as before, $x \in (l_1, l_2)$. The DPE for $J(x, \mu, \bar{u})$ is

$$0 = 1 - \alpha_1(x - \mu)^2 + \max_u \{\tfrac{1}{2}a(x, u)J_{xx} + b(x, u)J_x - \alpha_2(u - \bar{u})^2\}. \quad (6.5.14)$$

Exercise 6.5.4

Derive equation (6.5.14).

The idea behind adding the explicit cost term in the control is to move the boundary optimal control to the interior of the region. Then, for example, the method of Arkin et al. (1966) can be used to determine the controls. For simplicity, assume that the infinitesimal variance is independent of u, that is, that $a = a(x)$ only. Then assuming an internal maximum gives, from (6.5.14),

$$b_u(x, u)J_x - 2\alpha_2(u - \bar{u}) = 0. \quad (6.5.15)$$

Exercise 6.5.5

What conditions ensure that the maximum is internal?

The feedback control satisfies the equation

$$u(x) = [b_u(x, u)/2\alpha_2]J_x + \bar{u}. \quad (6.5.16)$$

Equation (6.5.16) is an implicit equation for $u(x)$, once $J(x, \mu, \bar{u})$ is known. Alternately, solving (6.5.15) for J_x gives

$$J_x = 2\alpha_2(u - \bar{u})/b_u(x, u) \equiv g_1(x, u), \quad (6.5.17)$$

and differentiating (6.5.17) gives, after some simplification,

$$J_{xx} = \frac{2\alpha_2 u'(x) - 2\alpha_2(u(x) - \bar{u})[b_{ux} + b_{uu}u'(x)]}{b_u(x, u)^2} \quad (6.5.18)$$

$$\equiv g_2(x, u, u'(x)). \quad (6.5.19)$$

Using (6.5.17), (6.5.18), and (6.5.19) in (6.5.14) converts the DPE to an ordinary differential equation for $u(x)$. This equation is

$$\tfrac{1}{2}a(x)g_2(x, u(x), u'(x)) + 1 - \alpha_1(x - \mu)^2$$
$$+ b(x, u)g_1(x, u(x)) - \alpha_2[u(x) - \bar{u}]^2 = 0. \quad (6.5.20)$$

Since $T(x)$ vanishes at either end of the interval (l_1, l_2), it must be that $J(l_1, \mu, \bar{u}) = J(l_2, \mu, \bar{u}) = 0$. Assume that (6.5.20) has the solution $u(x, u_0)$, where u_0 is the constant of integration. In light of (6.5.17), $J(x, \mu, \bar{u})$ can be written as

$$J(x, \mu, \bar{u}) = \int_{l_1}^{x} \frac{2\alpha_2(u(s, u_0) - \bar{u})}{b_u(s, u(s, u_0))} ds. \tag{6.5.21}$$

The constant u_0 is determined from the implicit equation

$$J(l_2, \mu, \bar{u}) = 0 = \int_{l_1}^{l_2} \frac{2\alpha_2(u(s, u_0) - \bar{u})}{b_u(s, u(s, u_0))} ds. \tag{6.5.22}$$

Exercise 6.5.6

Apply this theory to a population growth model with $b(x, u) = rx(1 - x) - ux$, $a(x, u) = a(x)$, and $\alpha_1 = 0$.

Exercise 6.5.7

In order to use this theory, μ and $\bar{u}$ are needed. How could they be determined?

Exercise 6.5.8

Since this theory was motivated by the idea of smoothing out control and stock fluctuations, one can assume that α_1 and α_2 are large. What simplifications arise in (6.5.20–22) if one assumes that α_1 and α_2 are large?

Exercise 6.5.9

Equation (6.5.20) is, in general, a nonlinear equation. How can one characterize the size of the region on which it is valid?

Bibliographic Notes

The approaches of Ludwig and Mendelssohn described in the the early part of this section require considerable numerical work in order to obtain solutions. An introduction to the numerical methods is given by Ludwig (1979b). Mendelssohn (1979, 1980) also discusses the numerical considerations. The concept of Pareto optimal controls is very useful in many economic settings. An introduction to such ideas from the economic perspective is found in Hirshleifer (1970) and from the control perspective in Leitmann (1981). Numerical methods for efficiently computing Pareto optimal policies are discussed in White and Kim (1980).

7

Management of Mixed Renewable and Exhaustible Resource Systems: Agricultural Pest Control

In agricultural pest control one is faced with the management of a mixed resource system. The crop under consideration is a renewable resource since it grows within the season according to some stock dynamics that depend upon biomass. It is assumed that there is a pest in the field and that the pest is sprayed with pesticide. It is commonly observed that, as pests are sprayed with pesticide, resistance to the pesticide develops. Consequently, one can view the susceptibility of the agricultural pest to pesticide as an exhaustible resource [see, e.g., Hueth and Regev (1974)]. By spraying a field heavily in a given year, a farmer potentially increases resistance to the pesticide, which, in principle, reduces yields in future years. The return to the farmer comes from the renewable resource, but it is the state of the exhaustible resource (in this case susceptibility to pesticide) that crucially affects the overall return. Consequently, the management problem involves the optimal use of both an exhaustible resource and a renewable resource. Many problems in agriculture can be viewed in this way. Consider, for example, the study of irrigation problems. In this case the water used for irrigation can be viewed as the exhaustible resource which is used to produce a return from the renewable resource. A problem in which two different exhaustible resources are involved is discussed by Epple and Lave (1982). In that case helium is lost (permanently) as natural gas is produced.

The question of interest is then, How does one conserve the helium while exploiting the natural gas?

The first section of this chapter contains a discussion of one operational model associated with the management problem for agricultural pest control. The second section contains models for the development of the biological populations, in the absence of any genetic models. Genetic models are needed in order to characterize the buildup of resistance to the pesticide. These models are introduced in the third section. By coupling genetics and population dynamics, the dynamics of the development of resistance can be studied. The techniques of dynamic programming and control theory can then be applied to the problem of resistance management. There is a vast literature on agricultural pest management. This chapter is only meant as an introduction to some of the problems. The references given in this chapter are not inclusive in any sense. Some good introductions to the literature are Comins (1977a, b), Hueth and Regev (1974), Georghiou (1980), Georghiou and Taylor (1977a, b), Taylor and Georghiou (1979, 1982), Watson, Moore, and Ware (1975), Varley, Gradwell, and Hassell (1973), Hassell (1978), and Charlesworth (1980). The material presented in this chapter is based on the work of Mangel and Plant (1983) and Plant, Mangel, and Flynn (1984), where further details can be found. The paper of Moffitt and Farnsworth (1981) treats a similar problem, but from a different viewpoint and with different methods.

7.1 THE OPERATIONAL FORMULATION: COTTON IN CALIFORNIA

The operational picture is the following one (it is based on the way that cotton is grown in the San Joaquin Valley in California). At the start of the agricultural season, there are no pests in the field. This occurs, for example, because the pest is crop specific and the crops are rotated among fields. Thus, it is assumed that the crop that was grown in the previous season did not support the pest. The host crop is planted and starts growing. Its growth is characterized by a dynamical equation for the biomass of the crop at time t, $C(t)$. For example, the paper of Gutierrez et al. (1975) contains a detailed description of a model for the growth of cotton.

Although pests are not present in the field at the start of the season, it is assumed that they are supported by secondary hosts (e.g., patches of weeds) outside of the field. Pests arrive at the field from the outside source at a rate $I(t)$. In the following the source of pests will be called the pool. The arrival rate depends upon such factors as temperature, kind of crop in the field,

7.1. THE OPERATIONAL FORMULATION

Table 7.1.1

Increase in LD_{50} (mg/g) for Tobacco Budworm[a]

Pesticide	Year				
	1961	1962	1963	1964	1964
Endrin	0.06	0.12	0.20	—	12.9
Carbyl	0.30	0.60	15.6	12.5	54.5

[a] From Adkisson (1968).

habitability of the pool, etc. In principle, one can determine $I(t)$ by measurements throughout the season, so that one can view $I(t)$ as a statistically known quantity. Thus, in the dynamic equations for the pest population, $I(t)$ can be viewed as a stochastic term.

As pests arrive they start colonies which grow over time. It is assumed that the colonies do not interact with each other. This assumption requires, of course, that the pests are sessile once they arrive. These assumptions are satisfied, for example, by colonies of spider mites, which are pests of cotton. If the pesticide is age specific, then the age structure of the colonies is important. In this case it will be seen that it is possible to use the density of ages within a colony as a state variable (since there is no interaction between the sessile colonies). If age structure is not important, then one can use the total number of pests as a state variable.

At some point within the season, the farmer sprays the field with pesticide. The spray does two things. First, it kills some fraction of the pests. Second, it provides a selection mechanism for any resistant genes that may be present in the population. Common measures for the effectiveness of a pesticide are the doses needed to kill 50% or 95% of the population. These are typically denoted by LD_{50} and LD_{95}, respectively. By tracking the values of either lethal dose, one can also track the development of resistance to the pesticide. For example, Table 7.1.1 shows the development of resistance to endrin and carbyl in the tobacco budworm. The development of resistance is measured through the increase in LD_{50}.

Bibliographic Notes

The operational picture developed in this section is based on cotton as the crop and spider mites as the pest. Cotton is one of the most heavily sprayed crops in the United States [see, e.g., Cooke and Parvin (1981)]. The spider mite is a good pest to study because at the present time there is no commonly implemented detection program for spider mites. Consequently, spraying is done according to a schedule. Background material on spider mites is found in Carey (1982a, b), Carey and Bradley (1982), Helle and Van Zon (1967), Wilson *et al.* (1983), and Bartlett (1968).

7.2 MODELS FOR PEST POPULATION DYNAMICS

The simplest case that can be envisioned is one in which all of the pests are equally susceptible to the pesticide. This allows one to concentrate on the development of the model for the pest population dynamics. According to the operational picture described in the previous section, as the pests arrive they establish colonies that age. Since there is no interaction between the colonies, it is possible to work directly with an age density within the colonies.

Let $x(t, a; n) \, da$ denote the number of pests in year n at time t in colonies of ages a to $a + \Delta a$. Thus, $x(t, a; n)$ is an age density for pest colonies. The total number of pests present in year n at intraseason time t, $X(t; n)$, is then found by integrating the density over all ages. This gives

$$X(t; n) = \int_0^t x(t, a; n) \, da. \qquad (7.2.1)$$

The integral only goes to t since assumptions preclude there being at time t any pests in colonies of age greater than t. Often in the following $x(t, a)$ and $X(t)$ will be used for $x(t, a; n)$ and $X(t; n)$.

Assume that the pests in the field grow at a rate $r(s(t), a)x(t, a)$, where $s(t)$ is the intensity of spray at time t. With these assumptions, compare $x(t + h, a + h)$ and $x(t, a)$. This comparison gives

$$x(t + h, a + h) - x(t, a) = r(s(t), a)x(t, a)h + o(h). \qquad (7.2.2)$$

Adding and subtracting $x(t, a + h)$ to the left-hand side, dividing by h, and letting $h \to 0$ gives the first-order equation

$$\partial x/\partial t + \partial x/\partial a = r(s(t), a)x(t, a). \qquad (7.2.3)$$

The right-hand side of (7.2.3) is linear in x because of the assumption about the growth rate of the pest. In general, one could assume that the pests grow at a rate $f(x(t, a), X(t), s(t))$, where f is a density-dependent growth function. The right-hand side of (7.2.3) is then replaced by $f(x(t, a), X(t), s(t))$.

Exercise 7.2.1

By what equation is (7.2.3) replaced if the growth of $x(t, a)$ is a logistic function with a carrying capacity K that pertains to the entire population?

Equation (7.2.3) is a partial differential equation in two variables. It requires boundary or initial conditions. What sort of boundary or initial conditions are needed? By assumption, there are no pests present at time 0. Thus $x(0, a) = 0$ for all a. In fact, it must be that $x(t, a) = 0$ if $a > t$. Now consider $x(t, 0)$. Since pests arrive at a rate $I(t)$,

$x(t, 0) \Delta t$ = number of pests in colonies of age 0 to Δt at time t

$$= \int_{t-\Delta t}^{t} I(s)\, ds + o(\Delta t). \tag{7.2.4}$$

Taylor expanding in (7.2.4), dividing by Δt, and letting $\Delta t \to 0$ gives the condition

$$x(t, 0) = I(t). \tag{7.2.5}$$

Thus, if $I(t)$ is stochastic, the situation is that a stochastic boundary condition arises when solving the partial differential equation (7.2.3).

The growth rate r in (7.2.3) depends upon the intensity of the pesticide since when the pesticide is applied it effectively reduces the intrinsic growth rate of the population. The particular form of the growth function is not important at this level of modeling. A convenient choice for $r(s(t), a)$ is

$$r(s(t), a) = r_0 \left\{ 1 - \frac{w(a)s(t)}{\varepsilon + s(t)} \right\}. \tag{7.2.6}$$

In this formula r_0 is the intrinsic growth rate in the absence of spraying, $w(a)$ measures how susceptible colonies of age a are to the spraying, and ε measures how quickly the spraying effect saturates. It is connected, for example, to the lethal dose levels. The particular form of $w(a)$ depends on the pest and pesticide. If $w(a)$ increases with a, then older pests are more susceptible to pesticide than young ones. Conversely, if $w(a)$ decreases with a, then the young pests are most susceptible.

The model (7.2.6) is completely deterministic. One can introduce uncertainty into the effect of pesticide on the pests by replacing the term $w(a)s(t)/[\varepsilon + s(t)]$ by a random function of $w(a)s(t)/[\varepsilon + s(t)]$ or, more simply, by a random function of $s(t)$. For example, one could assume that the logarithm of the fraction of pests killed is normally distributed with a mean that depends on the intensity of the application of the pesticide and a variance independent of the intensity.

For the actual form of $s(t)$, the following model is adopted. Given a spray initiation time $t_s(n)$ in year n and a length of activity of pesticide δ, assume that $s(t)$ in year n is given by

$$s(t) = \begin{cases} 0, & 0 \le t < t_s(n), \\ \eta(n), & t_s(n) \le t \le t_s(n) + \delta(n), \\ 0, & t > t_s(n) + \delta(n). \end{cases} \quad (7.2.7)$$

Equation (7.2.7) has the following interpretation. Only one spray is applied each year. The spray takes a pulse form, with constant intensity $\eta(n)$ over a fixed time period $\delta(n)$. If one thinks about the control problem, then the control variables would probably be $\eta(n)$ and $t_s(n)$, the intensity of spray and the time at which spraying commences. The value of $\delta(n)$ is most likely determined by exogenous factors, such as environmental conditions. Consequently, $\delta(n)$ could be viewed as a stochastic term as well.

Exercise 7.2.2

How does one relate the LD_{50}, which is a common measure of pesticide intensity, to $\eta(n)$, the measure of intensity used in this model?

Once $x(t, a)$ is known, $X(t)$ can be found by integration as in (7.2.1). If $w(a) = w$, a constant, (7.2.3) becomes

$$\partial x/\partial t + \partial x/\partial a = r(s(t))x. \quad (7.2.8)$$

In this case (7.2.8) integrates directly to give

$$\int_0^t \frac{\partial x}{\partial t} da + \int_0^t \frac{\partial x}{\partial a} da = r(s(t)) \int_0^t x(t, a)\, da, \quad (7.2.9)$$

or

$$dX/dt + x(t, t) - x(t, 0) = r(s(t))X(t). \quad (7.2.10)$$

Now $x(t, t) \equiv 0$ and since $x(t, 0) = I(t)$, one obtains

$$dX/dt = r(s(t))X(t) + I(t), \quad (7.2.11)$$

and the initial condition for the ordinary differential equation (7.2.11) is $X(0) = 0$. Equation (7.2.11) motivates the non-age-structured models presented below. Observe that if (7.2.11) is appropriate, then $I(t)$ is simply a stochastic forcing term and (7.2.11) is a SDE similar to the ones studied in previous chapters.

Exercise 7.2.3

Interpret the two terms on the right-hand side of (7.2.11).

7.2. MODELS FOR PEST POPULATION DYNAMICS

Exercise 7.2.4

Compute the solution of (7.2.11). Set $R(t) = \int_0^t r(s(t')) \, dt'$.

If $w(a)$ is not constant, the partial differential equation to be solved is

$$\frac{\partial x}{\partial t} + \frac{\partial x}{\partial a} = r_0 \left(1 - \frac{w(a)s(t)}{\varepsilon + s(t)}\right) x. \qquad (7.2.12)$$

Equation (7.2.12) is a first-order, linear partial differential equation with nonconstant coefficients. It can be solved by the method of characteristics (John, 1978). Consider three time regimes: $0 \le t < t_s$, $t_s < t \le t_s + \delta$, and $t_s + \delta < t \le T$. In the first regime (7.2.12) becomes

$$\partial x/\partial t + \partial x/\partial a = r_0 x. \qquad (7.2.13)$$

To use the method of characteristics, observe that a in (7.2.13) can be thought of as a function $a(t)$. If $a(t)$ satisfies

$$da/dt = 1, \qquad (7.2.14)$$

then (7.2.13) becomes

$$\frac{\partial x}{\partial t} + \frac{\partial x}{\partial a}\frac{da}{dt} = r_0 x. \qquad (7.2.15)$$

The left-hand side of (7.2.15) is the total derivative of $x(t, a(t))$. That is, (7.2.15) can be rewritten as

$$dx/dt = r_0 x, \qquad (7.2.16)$$

where (7.2.16) holds on the solution curves of (7.2.14).

The solutions of (7.2.14) and (7.2.16) are

$$a = t - c_1, \qquad x = c_2 e^{r_0 t}, \qquad (7.2.17)$$

where c_1 and c_2 are constants of integration.

Exercise 7.2.5

What is the interpretation of c_1 in terms of the operational picture?

Combining the two equations in (7.2.17) gives

$$x(t, a) = c(t - a)e^{r_0 t}, \qquad (7.2.18)$$

where the function $c(t - a)$ is to be determined. Since $x(0, a) = 0$, $c(-a) = 0$, which implies that $c \equiv 0$ for negative arguments, so that $x(t, a) \equiv 0$ if $a > t$. Applying the boundary condition (7.2.5) gives $c(t)e^{r_0 t} = I(t)$. Thus, $c(t) = I(t)e^{-r_0 t}$ and the solution is

$$x(t, a) = I(t - a)e^{r_0 a}. \qquad (7.2.19)$$

By assumption $I(s) \equiv 0$ if $s < 0$, so that the condition $x(t, a) = 0$ if $a > t$ is automatically satisfied. Equation (7.2.19) shows that $x(t, a)$ is proportional to $I(t)$ in the region $0 \leq t \leq t_s$. Thus, the statistical properties of $x(t, a)$ are easily and directly computed from (7.2.19).

Exercise 7.2.6

Interpret the solution (7.2.19) in terms of the operational picture.

At $t = t_s$ the differential equation becomes

$$\frac{\partial x}{\partial t} + \frac{\partial x}{\partial a} = r_0\left(1 - \frac{w(a)\eta}{\varepsilon + \eta}\right)x \qquad (7.2.20)$$

with boundary conditions $x(t_s, a) = I(t_s - a)e^{r_0 a}$ and (7.2.5).

Exercise 7.2.7

Find the solution of (7.2.12) in the regimes $(t_s, t_s + \delta)$ and $(t_s + \delta, T)$.

The complete solution of (7.2.12) is not sufficient for the problem of resistance management, since the development of resistance (or loss of susceptibility to pesticide) is not included in these models. In order to include the development of resistance, one must add genetic components to the models.

Bibliographic Notes

The book by Watson, Moore, and Ware (1975) provides a good introduction to the lethality of pesticides. They give many examples of LD_{50} for specific and broad-spectrum insecticides. In addition, they provide a good introduction to the literature of pesticides. Models of age-structured populations are discussed by Charlesworth (1980), Hoppensteadt (1975), and Maynard Smith (1974). The method of characteristics, a major tool of applied mathematics, is discussed in Courant and Hilbert (1962) and John (1978). The book by Aris and Amundson (1973) provides a multitude of examples of how the method is used in chemical problems.

7.3 MODELS WITH POPULATION GENETICS

The next step is to incorporate population genetics into the model. The simplest case is the one in which a single locus controls resistance and each individual carries two alleles—either resistant (R) or susceptible (S). With this picture there are two kinds of homozygous pests (allele combinations RR and SS) and one kind of heterozygous pest (RS). To show how genetics

7.3. MODELS WITH POPULATION GENETICS

can be incorporated, it is easiest to begin with a model that has no age structure. The variables of interest are then

$X_1(t; n)$ = number of homozygous resistant (RR) pests in the field at time t in season n,

$X_2(t; n)$ = number of heterozygous resistant (RS) pests in the field at time t in season n, (7.3.1)

$X_3(t; n)$ = number of homozygous susceptible (SS) pests in the field at time t in season n.

The resistance to pesticide can be characterized by assuming that, if the population is sprayed, the fraction of resistant pests that survives is larger than the fraction of susceptible pests that survives. In the absence of spraying, it is often observed that the susceptible pests are more fit than resistant pests. These two ideas are captured as follows. Let $\mu_i(n)$ denote the fraction of pests of class i (i = RR, RS, or SS) in the pool in year n. The rate of increase of pests of class i in the field due to influx from the pool is then $\mu_i(n)I(t)$. Assume that the $X_i(t; n)$ satisfy equations analogous to (7.2.11). These equations are

$$\frac{d}{dt} X_1(t; n) = r_1(s)X_1 + \mu_1(n)I + \frac{1}{2}\alpha X_2,$$

$$\frac{d}{dt} X_2(t; n) = r_2(x)X_2 + \mu_2(n)I - \alpha X_2, \quad (7.3.2)$$

$$\frac{d}{dt} X_3(t; n) = r_3(x)X_3 + \mu_3(n)I + \frac{1}{2}\alpha X_2.$$

Both the level of susceptibility and the fitness in the absence of spraying are captured in the growth terms $r_i(t)$. They are assumed to take the form

$$r_1(s) = r_{10}[1 - ws/(\varepsilon_1 + s)],$$

$$r_2(s) = r_{20}[1 - ws/(\varepsilon_1 + s)], \quad (7.3.3)$$

$$r_3(s) = r_{30}[1 - ws/(\varepsilon_2 + s)].$$

In (7.3.3) assume that $r_{30} > r_{10}$, so that pests carrying the susceptibility gene are more fit in the absence of spraying. Also, assume that $0 \leq \varepsilon_2 \ll \varepsilon_1$, so that spraying affects the susceptible pests long before the resistant ones. Since ε_1 appears in the growth rate $r_2(s)$, it is assumed that heterozygote pests are resistant (i.e., that R dominates S). The relationship between r_{20} and r_{10} depends upon the relative fitness of the heterozygote pests.

The parameter α in (7.3.2) measures the rate at which heterozygotes disappear due to genetic mixing. It can be calculated once a specific genetic model is given (Crow and Kimura, 1970).

The proportions of pests of the various classes in the next year depend on the proportions at the time that mating occurs and on the mechanism of mating. If the matings are random, then the proportions can be calculated from the Hardy–Weinberg formula (Crow and Kimura, 1970).

Exercise 7.3.1

In this exercise the Hardy–Weinberg formula is derived. Let f_n be the frequency of the R allele in year n and A, B, and C be the surviving homozygous resistant, heterozygous, and homozygous susceptible individuals, respectively, after pesticide is applied. Assume random mixing between individuals. Show that the frequency of the R allele in the next generation is

$$f_{n+1} = \frac{[Af_n + B(1 - f_n)]f_n}{Af_n^2 + 2Bf_n(1 - f_n) + C(1 - f_n)^2}. \tag{7.3.4}$$

(*Hint*: Compute the frequency of all possible matings after spraying.) If f_n is small, derive an approximation to (7.3.4) by Taylor expanding the right-hand side.

Exercise 7.3.2

Assume that $r_{20} = r_{10}$ and that $\alpha \equiv 0$ in (7.3.3). What happens to the three-equation system (7.3.3)? Show how in this situation one needs only think of the evaluation of R and S alleles.

The goal here is to relate $\mu_i(n + 1)$ to $\mu_i(n)$. To do this, let T_w be the effective length of the winter and set

$$p_i(n) = \frac{X_i(T; n) \exp(r_{i0} T_w)}{\sum X_i(T; n) \exp(r_{i0} T_w)}, \tag{7.3.5}$$

so that $p_i(n)$ is the fraction of the population in the ith species kind at the end of the winter after season n. Then following the Hardy–Weinberg assumption gives

$$\mu_1(n + 1) = [p_1(n) + \tfrac{1}{2}p_2(n)]^2,$$
$$\mu_2(n + 1) = 2[p_1(n) + \tfrac{1}{2}p_2(n)][p_3(n) + \tfrac{1}{2}p_3(n)], \tag{7.3.6}$$
$$\mu_3(n + 1) = [p_3(n) + \tfrac{1}{2}p_2(n)]^2.$$

7.3. MODELS WITH POPULATION GENETICS

Exercise 7.3.3

Derive (7.3.6) from (7.3.4). Observe that (7.3.6) is a set of deterministic difference equations for the $\mu_i(n)$.

In order to write (7.3.6), it is assumed that there are no pests in the pool at the end of the season. Since the ultimate goal is to use these models for control of resistance, this assumption implies a conservative strategy. In light of the many uncertainties in a real agricultural system and the stochastic effects, such a conservative approach seems reasonable.

Exercise 7.3.4

Suppose that the pool is not exhausted at the end of the season. Show that the pool can be viewed as an additional source of susceptibility. How must (7.3.6) be modified?

The solutions of (7.3.2) and thus the difference relationships (7.3.6) are obtained in a straightforward manner; numerical results are presented in Mangel and Plant (1983) and Plant, Mangel, and Flynn (1984).

The next model incorporates age structure into a model with genetics. Again, using a single-locus, two-allele genetic model one sets

$$\begin{aligned}x_1(t, a; n) &= \text{number of homozygous resistant (RR) pests in} \\ &\quad \text{colonies of age } a \text{ at time } t \text{ in season } n, \\ x_2(t, a; n) &= \text{number of heterozygous resistant (RS) pests in} \\ &\quad \text{colonies of age } a \text{ at time } t \text{ in season } n, \\ x_3(t, a; n) &= \text{number of homozygous susceptible (SS) pests in} \\ &\quad \text{colonies of age } a \text{ at time } t \text{ in season } n.\end{aligned} \quad (7.3.7)$$

In analogy with (7.2.3), these variables are assumed to satisfy

$$\begin{aligned}\frac{\partial x_1}{\partial t} + \frac{\partial x_1}{\partial a} &= r_{10}\left[1 - \frac{w(a)s(t)}{\varepsilon_1 + s(t)}\right]x_1 + \frac{1}{2}\alpha x_2, \\ \frac{\partial x_2}{\partial t} + \frac{\partial x_2}{\partial a} &= r_{20}\left[1 - \frac{w(a)s(t)}{\varepsilon_1 + s(t)}\right]x_2 - \alpha x_2, \\ \frac{\partial x_3}{\partial t} + \frac{\partial x_3}{\partial a} &= r_{30}\left[1 - \frac{w(a)s(t)}{\varepsilon_2 + s(t)}\right]x_3 + \frac{1}{2}\alpha x_3,\end{aligned} \quad (7.3.8)$$

with the boundary conditions

$$x_i(0, a) = 0, \qquad x_i(t, 0) = \mu_i I. \quad (7.3.9)$$

Equations (7.3.8) are a system of first-order partial differential equations. In the period before spraying, direct solution gives

$$x_2(t, a) = \mu_2 I(t - a) \exp[(r_{20} - \alpha)a]. \tag{7.3.10}$$

Then $x_1(t, a)$ and $x_3(t, a)$ satisfy the equation

$$\frac{\partial x_i}{\partial t} + \frac{\partial x_i}{\partial a} = r_{i0} x_i + \frac{1}{2} \alpha \mu_2 I(t - a) \exp[(r_{20} - \alpha)a], \quad i = 1 \text{ or } 3. \tag{7.3.11}$$

Exercise 7.3.5

Obtain the solutions of (7.3.11) in the period $(0, t_s)$.

Exercise 7.3.6

Complete the characterization of the solutions of (7.3.8) in $(t_s, t_s + \delta)$ and $(t_s + \delta, T)$.

Bibliographic Notes

The literature on the theory of population genetics is considerable. Good introductions, of increasing complexity, are found in the books by Maynard Smith (1968), Roughgarden (1979), Crow and Kimura (1970), and Waltman (1983). There are many uses for stochastic differential equations in population genetics. Again Crow and Kimura (1970), as well as Ludwig (1974, 1975) provide good introductions to the applications of SDEs in genetic theory.

7.4 THE SINGLE-SEASON ECONOMIC OPTIMIZATION PROBLEM

The development of resistance is of most interest in the context of the damage done to the crop by the pest. In order to study the effect of the pest on the crop, one needs a model for the crop–pest interaction. For the crop dynamics, adopt the model

$$\frac{dC}{dt} = \begin{cases} r_c C - v(X_1(t) + X_2(t) + X_3(t)), & C > 0, \ 0 \le t < T, \\ 0, & \text{otherwise.} \end{cases} \tag{7.4.1}$$

Here r_c is the intrinsic growth rate of the crop and v represents the rate at which damage is done to the crop per pest. This equation represents the simplest model for damage.

7.4. SINGLE-SEASON ECONOMIC OPTIMIZATION

Exercise 7.4.1

Formulate models for the crop dynamics that include

(i) density dependence upon the crop,
(ii) density dependence upon the pest, and
(iii) an age-specific damage function.

In most agricultural situations, damage cannot be measured incrementally, but should be measured in terms of reduced yield at the end of the growing season.

Consider the problem of optimizing economic gain and susceptibility at the end of a single growing season of length T. Suppose that $C(T)$ denotes the biomass of crop at the end of the season, p the price per unit biomass of crop of received at harvest, and c_s the cost per unit time of spraying. A reasonable objective functional is then

$$J = E\left\{\gamma pC(T) + (1-\gamma)p'\left[\frac{X_3(T) + \tfrac{1}{2}X_2(T)}{X_1(T) + X_2(T) + X_3(T)}\right]^2 - \int_0^T c_s s(t)\,dt\right\}. \tag{7.4.2}$$

In this equation $0 \le \gamma \le 1$ and p' is the economic value of susceptibility; it depends, of course, on future economic considerations, but for the simple problem here take it as a fixed, known value. A way of determining p' is presented in the next section. The first term in (7.4.2) is the value of the crop yield, weighted by γ. The second term in (7.4.2) is the value of the remaining susceptibility in the population, and the third term is the cost of applying pesticide. The problem now is to maximize (7.4.2).

In the case of a population without age structure, the single-season optimization problem takes the form

$$\max E\left\{\gamma pC(T) + (1-\gamma)p'\left[\frac{X_3(T) + \tfrac{1}{2}X_2(T)}{X_1(T) + X_2(T) + X_3(T)}\right]^2 - \int_0^T c_s s(t)\,dt\right\} \tag{7.4.3}$$

such that

$$\dot{C} = r_c C - v(X_1(t) + X_2(t) + X_3(t)),$$
$$\dot{X}_2 = r_2(s(t))X_2 - \alpha X_2 + \mu_2 I(t), \tag{7.4.4}$$
$$\dot{X}_i = r_i(s(t))X_i + \tfrac{1}{2}\alpha X_2 + \mu_i I(t), \quad i = 1 \text{ or } 3.$$

If $s(t)$ has the pulse form described in (7.2.7), one must choose t_s and η in an optimal fashion. This is a parameter optimization problem; the cost term in the functional simplifies to $-c_s \eta \delta$ and t_s enters the problem via the differential equations only.

Exercise 7.4.2

Solve the equations (7.4.4), substitute into (7.4.3), and thus obtain an explicit form for the objective functional. Characterize the optimal choices of η and t_s in terms of the statistical properties of $I(t)$ and δ. How do these results simplify if one uses a deterministic model?

If the pulse form is not required for $s(t)$ and $I(t)$ is treated as a deterministic function, then the optimal choice for $s(t)$ can be determined by control theory. This is useful to do because it shows how the optimal spraying strategy depends upon the various biological parameters. Introduce adjoint variables λ_0 corresponding to C and λ_i corresponding to X_i. The Hamiltonian for the problem is

$$\mathcal{H} = -c_s s(t) + \lambda_0 [r_c C - v_1(X_1(t) + X_2(t) + X_3(t))]$$
$$+ \lambda_2 [r_2(s(t))X_2 - \alpha X_2 + \mu_2 I(t)]$$
$$+ \sum_{i=1,3} \lambda_i [r_i(s(t))X_i + \tfrac{1}{2}\alpha X_2 + \mu_i I(t)]. \tag{7.4.5}$$

In the following it helps to set

$$\varphi = \frac{[X_3(T) + \tfrac{1}{2}X_2(T)]^2}{[X_1(T) + X_2(T) + X_3(T)]^2}(1-\gamma)p'. \tag{7.4.6}$$

The adjoint variables satisfy the equations

$$\dot{\lambda}_0 = -\partial \mathcal{H}/\partial C = -\lambda_0 r_c, \qquad \lambda_0(T) = \gamma p; \tag{7.4.7}$$

$$\dot{\lambda}_2 = -\partial \mathcal{H}/\partial X_2 = v\lambda_0 - \lambda_2[r_2(s(t)) - \alpha] - \tfrac{1}{2}\alpha(\lambda_3 + \lambda_1),$$
$$X_2(T) \geq 0, \qquad \lambda_2(T) \geq \partial \varphi/\partial X_2,$$
$$X_2(T)[\lambda_2(T) - \partial \varphi/\partial X_2] = 0; \tag{7.4.8}$$

$$\dot{\lambda}_1 = -\partial \mathcal{H}/\partial X_1 = v\lambda_0 - \lambda_1 r_1(s(t)),$$
$$X_1(T) \geq 0, \qquad \lambda_1(T) \geq \partial \varphi/\partial X_1,$$
$$X_1(T)[\lambda_1(T) - \partial \varphi/\partial X_1] = 0; \tag{7.4.9}$$

$$\dot{\lambda}_3 = -\partial \mathcal{H}/\partial X_1 = v\lambda_0 - \lambda_3 r_3(s(t)),$$
$$X_3(T) \geq 0, \qquad \lambda_3(T) \geq \partial \varphi/\partial X_3,$$
$$X_3(T)[\lambda_3(T) - \partial \varphi/\partial X_3] = 0. \tag{7.4.10}$$

Exercise 7.4.3

Derive the equations for the adjoint variables as well as the end conditions. Under what conditions could they be simplified?

7.4. SINGLE-SEASON ECONOMIC OPTIMIZATION

The maximum principle requires that the optimal $s(t)$ maximizes the Hamiltonian. Write $\mathcal{H}$ as

$$\mathcal{H} = \mathcal{H}_1 - c_1 s(t) - \lambda_2 r_{20}\left(\frac{ws(t)}{\varepsilon_1 + s(t)}\right) X_2$$

$$- \lambda_3 r_{30}\left(\frac{s(t)}{\varepsilon_2 + s(t)}\right) X_3 - \lambda_1 r_{10}\left(\frac{ws(t)}{\varepsilon_1 + s(t)}\right) X_1, \quad (7.4.11)$$

where $\mathcal{H}_1$ contains all terms independent of $s(t)$. Differentiating $\mathcal{H}$ gives

$$\frac{\partial \mathcal{H}}{\partial s} = -c_1 - \lambda_2 r_{20} \frac{w\varepsilon_1}{[\varepsilon_1 + s(t)]^2} X_2$$

$$- \lambda_3 r_{30} \frac{w\varepsilon_2 X_3}{[\varepsilon_2 + s(t)]^2} - \lambda_1 r_{10} \frac{w\varepsilon_1 X_1}{[\varepsilon_1 + s(t)]^2}. \quad (7.4.12)$$

Assuming an internal maximum and setting $\partial \mathcal{H}/\partial s = 0$ gives a fourth-order equation for $s(t)$; the solutions of interest are those that maximize $\mathcal{H}$.

Exercise 7.4.4

Characterize the solutions of this fourth-order equation. What happens if $\varepsilon_1 \to \infty$ (so that the resistant gene is perfectly resistant)?

Exercise 7.4.5

Solve equations (7.4.6) to (7.4.10) and characterize the optimal single-season spraying strategy.

Exercise 7.4.6

The results (7.4.5)–(7.4.12) require the assumption that $I(t)$ is deterministic. If it is not, then control theory cannot be used. However, one can still solve (7.4.4) and use the explicit formulation of (7.4.3) to characterize the optimal spraying strategy. Do this for the case in which $\alpha = 0$.

The results presented here provide a complete solution of the deterministic single-season spraying problem. To truly assess the effects of resistance, rather than introducing them simply as a parameter [as in (7.4.3)], one must proceed to the multiseason problem.

Bibliographic Notes

The model for the crop–pest interaction discussed here is a mere caricature of reality. Considerable effort must be expended to obtain more realistic models. [See, for example, Gutierrez et al. (1975) for a more thorough modeling effort.] The choice of an appropriate decision

7.5 THE MULTISEASON ECONOMIC OPTIMIZATION PROBLEM

The truly interesting optimization problem is the management of resistance over many seasons, since pesticide applied in a given year affects the future economic yields through reduced susceptibility to pesticide.

Assume that the time horizon for the management problem is N seasons. The decision-maker wishes to maximize the crop harvest over N seasons, subject to future discounting. Assume that the decision-maker expects a new pest control technology to be available at the time horizon, so that pest resistance to the pesticide does not enter explicitly into the cost function. This function may therefore be written as

$$J = E\left\{\sum_{n=0}^{N} \rho^n [C(T; n) - c_p \eta(n)]\right\}. \tag{7.5.1}$$

Here $C(T; n)$ is the solution at $t = T$ in the nth season of (7.4.1), c_p the cost per unit of pesticide, and $\eta(n)$ the intensity of pesticide in year n.

Exercise 7.5.1

Formulate an objective functional that depends on resistance to the pesticide.

Exercise 7.5.2

For the objective functional (7.5.1), what does heuristic reasoning suggest for the dependence of the control policy on the discount rate?

For the rest of this section, the pulse form for $s(t)$ is adopted, with $\eta(n)$ and $t_s(n)$ as the control variables. In addition, it will be assumed that $I(t)$ and δ are deterministic quantities, so that the dynamic programming equation for J will be a deterministic DPE.

7.5. MULTISEASON ECONOMIC OPTIMIZATION

Table 7.5.1

Value of the Objective Functional with Stationary Control Policies[a]

$\eta(n)$	$t_s(n)$	J
2	6	32.1
1.5	6	30.4
2.5	6	33.2
2	4	31.2
2	8	29.0

[a] The results in this table correspond to the following values of the parameters: $r_c = 1$, $v = 0.005$, $T = 20$, $\alpha = 1$, $\varepsilon_1 = 35$, $\varepsilon_2 = 5$, $r_{10} = 0.2$, $r_{20} = r_{30} = 0.21$, $\delta = 2$, $\eta = 2$, $w(a) = 20$, $\rho = 0.9$, $\mu_1(0) = 10^{-8}$, $M = 10$, and $C_0 = 1$. Further results are given in the papers of Mangel and Plant (1983) and Plant, Mangel, and Flynn (1984).

Based on the results of Sections 7.2 and 7.3, the dynamic variable for this problem can be viewed as f_n, the frequency of allele R in generation n. From Section 7.3, f_n satisfies a difference equation of the form

$$f_{n+1} = h(f_n, t_s(n), \eta(n)), \qquad f_1 = f_1^0 \text{ given.} \tag{7.5.2}$$

Exercise 7.5.3

Describe the algorithm used to calculate $h(f_n, t_s(n), \eta(n))$.

If $J_m^*(\mu)$ is the maximum value of J when $N = m$ and $f_m = \mu$, the DPE for $J_m^*(\mu)$ is

$$J_{m+1}^*(\mu) = \max_{t_s, \eta} \{C(T, m) - c_p\eta + \rho J_m^*(h(\mu))\}. \tag{7.5.3}$$

Exercise 7.5.4

Derive (7.5.3).

Equation (7.5.3) is a deterministic DPE, but it is not much easier to solve than the stochastic ones considered in other chapters.

Numerical solutions of (7.5.3) are described in Plant, Mangel, and Flynn (1984). Table 7.5.1 shows some of the dependence of J when η and t_s are held fixed, so that the control policy is stationary.

7. MIXED RENEWABLE AND EXHAUSTIBLE RESOURCE SYSTEMS

Exercise 7.5.5

If $I(t)$ is not deterministic, then (7.5.2) must be replaced by a stochastic difference equation. What is the form of this difference equation? What is the DPE for J if f_n satisfies a stochastic difference equation?

Bibliographic Notes

A full solution of the DPE (7.5.3) is found in Plant, Mangel, and Flynn (1984). There are many references on deterministic dynamic programming. Some standards are Bellman (1957), Bellman and Dreyfus (1962), Dreyfus and Law (1977), and Jacobson and Mayne (1970).

8

Introduction to Numerical Techniques

The requirement for numerical methods in stochastic control should be apparent to the reader by now. In general, each problem presents its own special kind of numerical difficulties. In addition, it is difficult to find a "canned routine" for stochastic optimization or control. Consequently, it is most important to have a good sense of the ideas behind some of the numerical techniques. With this in mind, this chapter contains a discussion of the following useful numerical techniques: Newton's method, series acceleration methods, solving SDEs numerically, and numerical methods for DPEs. The goal here is to expose the reader to some of the basic ideas behind numerical methods, to present some of the jargon, and to point to the appropriate places in the literature.

A good undergraduate-level reference for some of these topics is Burden, Faires, and Reynolds (1978); a more advanced treatment is found in Blum (1972). Further bibliographic notes are given at the end of each section.

8.1 NEWTON'S METHOD

Almost every section in this book involves a nonlinear equation. Newton's method provides a means of solving nonlinear algebraic equations. It is first illustrated for a single nonlinear equation; then, systems of equations are considered.

To start, consider the simple equation

$$f(x) = 0. \tag{8.1.1}$$

Here $f(x)$ is a given function of a single variable and it is assumed that $f(x)$ is twice differentiable. Assume that x_t is the solution of (8.1.1) and that $f'(x_t) \neq 0$. A Taylor expansion of (8.1.1) around x_t gives

$$f(x_t) = 0 = f(x) + f'(x)(x_t - x) + O((x_t - x)^2). \tag{8.1.2}$$

Solving (8.1.2) for x_t gives

$$x_t = x - f(x)/f'(x) + O((x_t - x)^2). \tag{8.1.3}$$

Equation (8.1.3) suggests the following iterative scheme:

(1) set $x_0 = x$, an initial guess;
(2) set $x_{n+1} = x_n - f(x_n)/f'(x_n)$;   (8.1.4)
(3) either exit or go to step 2.

When implementing such a scheme, the decision to exit is typically determined by whether $f(x_n)$ is close enough to zero. This depends on the particular problem. Although the scheme (8.1.4) is based on the approximation (8.1.3), it often turns out to converge to x_t quite rapidly.

Exercise 8.1.1

When would this scheme be exact?

It often happens that evaluating the derivative in step 2 of the algorithm is difficult; in this case it is sometimes useful to replace $f'(x_n)$ in step 2 of the algorithm by a numerical approximation for the derivative. Thus, step 2 is replaced by

$$x_{n+1} = x_n - f(x_n) \left[\frac{x_n - x_{n-1}}{f(x_n) - f(x_{n-1})} \right]. \tag{8.1.5}$$

Naturally, two starting values are needed if (8.1.5) is used.

It is a standard exercise in numerical analysis to show that there exists a δ such that Newton's method generates a sequence $x_1, x_2, \ldots$ converging to x_t for any $x_0 \in [x_t - \delta, x_t + \delta]$. It is worthwhile to see how such a proof proceeds. The basic idea behind the proof is to form a function

$$g(x) = x - f(x)\varphi(x) \tag{8.1.6}$$

and pick $\varphi(x)$ so that $g(x)$ has a fixed point, that is, a point where $g(x) = x$. Then $g(x_t) = x_t$ implies $f(x_t) = 0$ if $\varphi(x_t) \neq 0$. Equations (8.1.2) and (8.1.3) motivate the choice $\varphi(x) = 1/f'(x)$. So consider the function $g(x)$ defined by

$$g(x) = x - f(x)/f'(x). \tag{8.1.7}$$

8.1. NEWTON'S METHOD

Before proceeding, some results from calculus are needed. These are the following.

(1) If $g(x)$ is continuous for $a \leq x \leq b$ and $a \leq g(x) \leq b$ for all $a \leq x \leq b$, then there is at least one fixed point where $g(x) = x$ in $[a, b]$. To see this note that if $g(a) = a$ or $g(b) = b$, the proof is finished. If not, $g(a) > a$ and $g(b) < b$. Form $h(x) = g(x) - x$; then $h(a) > 0$, $h(b) < 0$, and the intermediate value theorem shows that there is an x with $h(x) = 0$.

(2) If $g(x)$ is differentiable for $a \leq x \leq b$, $a \leq g(x) \leq b$ for all $a \leq x \leq b$, and $|g'(x)| \leq k < 1$ for all $a \leq x \leq b$, then the fixed point is unique. To see this, suppose that x_1 and x_2 are two fixed points. Then, by the mean value theorem, there is $\xi \in [a, b]$ such that

$$|x_1 - x_2| = |g(x_1) - g(x_2)| = |g'(\xi)||x_1 - x_2| \leq k|x_1 - x_2| < |x_1 - x_2|,$$

giving a contradiction.

(3) Suppose that $g(x)$ is differentiable for $a \leq x \leq b$, $a \leq g(x) \leq b$, and $|g'(x)| \leq k < 1$. Then, if x_0 is in $[a, b]$, the sequence given by $x_n = g(x_{n-1})$ converges to the fixed point of $g(x)$. To see this, let x_f be the fixed point. Then, $|x_n - x_f| = |g(x_{n-1}) - g(x_f)| \leq |g'(\xi)||x_{n-1} - x_f| < k|x_{n-1} - x_f|$, where $a \leq \xi \leq b$. Applying this argument inductively gives $|x_n - x_f| < k^n|x_0 - x_f| \to 0$ as $n \to \infty$.

These results are applied to (8.1.7) by finding an interval such that $g(x)$ takes $[x_t - \delta, x_t + \delta]$ into itself and $|g'(x)| < 1$ there. Then one uses the iteration scheme $x_n = g(x_{n-1})$.

Exercise 8.1.2

Before reading further, try to provide the details of the proof.

The details of the proof proceed as follows. By assumption, $f'(x_t) \neq 0$ and $f'(x)$ is continuous, so that there is $\delta_1 > 0$ such that $f'(x) \neq 0$ for

$$x \in [x_t - \delta_1, x_t + \delta_1].$$

Thus, $g(x)$ in (8.1.7) is defined and continuous on $[x_t - \delta_1, x_t + \delta_1]$. Differentiating (8.1.7) gives $g'(x) = f(x)f''(x)/[f'(x)]^2$, so that $g'(x_t) = 0$. This implies that there exists a $\delta < \delta_1$ such that $|g'(x)| < 1$ for $x \in [x_t - \delta, x_t + \delta]$. To show that $g(x)$ takes $[x_t - \delta, x_t + \delta]$ into itself, note that if s is in this interval, then $|g(s) - x_t| = |g(s) - g(x_t)| = |g'(\xi)||s - x_t| \leq k|s - x_t| < |s - x_t|$ for $s \leq \xi \leq x_t$. Since $s \in [x_t - \delta, x_t + \delta]$, $|s - x_t| < \delta$ and $|g(s) - x_t| < \delta$, proving that $g(x)$ takes the interval into itself. Applying result 3 shows that the scheme $x_n = g(x_{n-1})$ converges to the fixed point, that is, to the solution of $f(x) = 0$.

Exercise 8.1.3

Estimate the rate of convergence of the x_n to x_t.

This essentially completely describes Newton's method for a single equation and a single variable.

All of the above was warmup for the multidimensional theory. So now consider the problem of solving

$$f_1(x) = 0, \quad f_2(x) = 0, \ldots, f_n(x) = 0. \tag{8.1.8}$$

Here $x = (x_1, \ldots, x_n)$ is an n-dimensional vector and each $f_i(x)$ is a given function. Let $F(x) = (f_1(x), f_2(x), \ldots, f_n(x))^T$, so that (8.1.8) becomes

$$F(x) = 0. \tag{8.1.9}$$

Before proceeding, multidimensional versions for the three results from calculus are needed. To obtain them, let $G(x) = (g_1(x), \ldots, g_n(x))^T$ be a given vector function of x.

(4) Suppose that a domain D is specified by $D = \{x = (x_1, \ldots, x_n) : x_i \in [a_i, b_i], i = 1, \ldots, n\}$, where the a_i and b_i are known constants. If $G(x)$ is continuous with continuous partial derivatives in D and $G(x) \in D$ for all $x \in D$, then $G(x)$ has a fixed point $x_t \in D$, where $G(x_t) = x_t$. In addition, if there exists a constant k such that $|\partial g_i/\partial x_j| \leq k/n$ for all $x \in D$ and for all i, j, then the fixed point is unique.

Exercise 8.1.4

How would one prove this result? (*Hint*: consider the case $n = 2$ first).

Next, one needs an iteration scheme. The natural choice is $X^{m+1} = G(x^m)$. In order to analyze this scheme, introduce a norm $\|\cdot\|_\infty$ for n-dimensional vectors by $\|v\|_\infty = \sup_i |v_i|$. Then the following result holds.

(5) Suppose that $G(x)$ has all of the properties specified in result 4. For a given x^0, set $x^{m+1} = G(x^m)$. Then, the sequence x^m converges to x_t and $\|x^m - x_t\|_\infty \leq [k^n/(1-k)]\|x^1 - x^0\|_\infty$.

Exercise 8.1.5

How would one prove this result?

With these results, the development of the multidimensional version of Newton's method can be motivated. To start, suppose that one wants to solve $F(x) = 0$. Note that

$$G(x) = x - A(x)^{-1} F(x) \tag{8.1.10}$$

8.1. NEWTON'S METHOD

has a fixed point at the solution of $F(x) = 0$ as long as $A(x)^{-1}$ is defined there. Thus, (8.1.10) provides the iteration scheme, once $A(x)$ is known. To motivate a choice for $A(x)$, one more result is needed.

(6) Suppose that $G(x) = (g_1(x), \ldots, g_n(x))^T$ has a fixed point at x_t and that there exists a number $\delta > 0$ with the following properties:

(1) $\partial g_i/\partial x_j$ is continuous for $x \in N_\delta \equiv \{x: \|x - x_t\|_\infty < \delta\}$ and for all i, j;
(2) $\partial^2 g_i/\partial x_j \partial x_k$ is continuous and bounded for $x \in N_\delta$ and all i, j, k. Let M be this bounding constant.
(3) $\partial g_i/\partial x_j = 0$ when $x = x_t$ for all i, j.

Define the sequence $x^m = G(x^{m-1})$. Then, x^m converges to x_t for any x^0 in N_δ and $\|x^m - x_t\|_\infty \leq (n^2 M/2) \|x^{m-1} - x_t\|_\infty$ for $m \geq 1$.

Exercise 8.1.6

How would one prove this result?

Let $A(x) = (a_{ij})$ in (8.1.10) and let the inverse matrix $A^{-1}(x) = (a^{ij})$. Writing (8.1.10) in components gives

$$g_i(x) = x_i - \sum_{j=1}^n a^{ij}(x) f_j(x), \tag{8.1.11}$$

so that

$$\frac{\partial g_i}{\partial x_k} = \delta_{ik} - \sum_{j=1}^n a^{ij} \frac{\partial f_j}{\partial x_k} - \sum_{j=1}^n f_j \frac{\partial a^{ij}}{\partial x_k}. \tag{8.1.12}$$

Here $\delta_{ik} = 1$ if $i = k$ and 0 otherwise (the Kroenecker delta). Setting (8.1.12) equal to zero at x_t and then setting $i = k$ gives

$$1 = \sum_{j=1}^n a^{ij} \frac{\partial f_j}{\partial x_i}\bigg|_{x_t} \tag{8.1.13}$$

[since $f(x_t) \equiv 0$]. If $i \neq k$, then (8.1.12) becomes

$$0 = \sum_{j=1}^n a^{ij} \frac{\partial f_j}{\partial x_k}\bigg|_{x_t}. \tag{8.1.14}$$

If the Jacobian matrix J is defined by

$$J(x) = (\partial f_i(x)/\partial x_j) \tag{8.1.15}$$

and I is the identity matrix, then (8.1.13) and (8.1.14) imply that

$$A(x_t)^{-1} J(x_t) = I, \tag{8.1.16}$$

so it appears that

$$A(x_t) = J(x_t). \tag{8.1.17}$$

This motivates the general choice $A(x) = J(x)$. Putting all these results together gives the iteration scheme

$$G(x) = x - J(x)^{-1}F(x),$$
$$x^{(k+1)} = x^{(k)} - J(x^{(k)})^{-1}F(x^{(k)}). \qquad (8.1.18)$$

For this method to work, the Jacobian must be nonsingular; if it is singular the idea behind Newton's method can still be used but more computations are needed [see, e.g., Decker and Kelley (1980)].

Even if $J(x)$ is nonsingular, inverting it can be quite a chore. One way around this is to let $y^{(k)}$ satisfy the linear system

$$J(x^{(k)})y^{(k)} = -F(x^{(k)}). \qquad (8.1.19)$$

Once the solution of (8.1.19), $y^{(k)}$, is known, one writes $x^{(k+1)} = x^{(k)} + y^{(k)}$ and uses that in the iteration scheme.

Exercise 8.1.7

In addition to the difficulties with the Jacobian, what other difficulties are associated with this method?

Bibliographic Notes

The first part of this section (on the one-dimensional Newton method) comes from Burden, Faires, and Reynolds (1978). A good treatment of the multidimensional theory is found in the book by Ortega (1972).

8.2 SEQUENCE AND SERIES ACCELERATION TECHNIQUES

Newton's method generates a sequence $x^{(n)}$ converging to x_t, but the convergence may not be as quick as one would like. Even more important, in stochastic dynamic programming one is often faced with summing series of the form

$$R = \sum_{n=0}^{\infty} p_n J^*(n), \qquad (8.2.1)$$

where p_n is the probability of the nth occurrence (e.g., discovery of n schools of fish) and $J^*(n)$ is the optimal future expected return conditioned on this

8.2 SEQUENCE AND SERIES ACCELERATION TECHNIQUES

discovery. Typically, such sums converge slowly. To see how to accelerate such a convergence, replace (8.2.1) by the finite sum

$$R_N = \sum_{n=0}^{N} p_n J^*(n). \tag{8.2.2}$$

The goal is to study the convergence of R_N. In particular, one wants to know if there are ways to rewrite the sequence R_N, $N = 0, 1, 2, \ldots$ so that it converges more rapidly to the sum in (8.2.1).

One simple series acceleration technique is called Aitken's Δ^2 method. It is motivated as follows. Suppose that $R_N \to R$ and that if $e_N = R_N - R$, then

$$\lim_{N \to \infty} |e_{N+1}|/|e_N| = \lambda < 1. \tag{8.2.3}$$

Assume for the moment that all the e_N are positive; then, $e_{N+1} = \lambda e_N$ and

$$R_{N+1} = e_{N+2} + R = \lambda e_{N+1} + R = \lambda(R_{N+1} - R) + R. \tag{8.2.4}$$

Equation (8.2.4) can be rewritten to give a difference equation for R_N. It is

$$R_{N+1} = \lambda(R_N - R) + R. \tag{8.2.5}$$

Solving (8.2.4, 5) for R gives the equation

$$\begin{aligned}
R &= \frac{R_{N+2} R_N - R_{N+1}^2}{R_{N+2} - 2R_{N+1} + R_N} \\
&= \frac{R_N^2 + R_N R_{N+2} + 2R_N R_{N+1} - 2R_N R_{N+1} - R_N^2 - R_{N+1}^2}{R_{N+2} - 2R_{N+1} + R_N} \\
&= \frac{(R_N^2 + R_N R_{N+2} - 2R_N R_{N+1}) - (R_N^2 - 2R_N R_{N+1} + R_{N+1}^2)}{R_{N+2} - 2R_{N+1} + R_N} \\
&= R_N - \frac{(R_{N+1} - R_N)^2}{R_{N+2} - 2R_{N+1} + R_N}.
\end{aligned} \tag{8.2.6}$$

Although the assumption of positivity of errors may not hold, this calculation motivates a new series defined by

$$\hat{R}_N = R_N - \frac{(R_{N+1} - R_N)^2}{R_{N+2} - 2R_{N+1} + R_N}. \tag{8.2.7}$$

The series (8.2.7) should converge more rapidly to R than the original series.

Exercise 8.2.1

Suppose that $\delta_N = |R - R_N|$ and that $\delta_N \to 0$ as $N \to \infty$. Compute or bound $\hat{\delta}_N = |R - \hat{R}_N|$ and estimate the acceleration in convergence.

Example 8.2.2

A prototypical problem is the stochastic dynamic programming problem discussed in Chapter 2. Such problems involve sums of the form

$$R = \sum_{n=0}^{\infty} \frac{1}{n!} \frac{\Gamma(n+v_1)}{\Gamma(v_1)} \frac{\alpha_1^{v_1}}{(\alpha_1+1)^{v_1+n}} \max\left(\frac{v_1+n}{\alpha_1+1}, \frac{v_2}{\alpha_2}\right). \quad (8.2.8)$$

Here v_1, v_2, α_1, and α_2 are parameters. Then R_N is defined by a sum similar to (8.2.8) but running from $n = 0$ to $n = N$.

Exercise 8.2.3

Code the sum (8.2.8) for values of v_i and α_i in the range 0.1 to 1 and experiment with the efficacy of the series acceleration method.

Bibliographic Notes

Burden, Faires, and Reynolds (1978) discuss a number of series acceleration techniques in addition to Aitken's Δ^2 method. The other techniques require the evaluation of derivatives. Bender and Orszag (1978) have an excellent chapter on the summation of series in which a variety of series acceleration techniques are discussed.

8.3 SOLVING STOCHASTIC DIFFERENTIAL EQUATIONS NUMERICALLY

One approach to studying an SDE of the form

$$dX = b(X, t)\, dt + \sqrt{a(X, t)}\, dW \quad (8.3.1)$$

is to solve it numerically. Here dW is standard Brownian motion with $E\{dW^2\} = dt$. This section contains a discussion of the problem of constructing a numerical solution of (8.3.1). Readable introductions are the papers by Helfand (1979) and Rao, Borwankar, and Ramkrishna (1974). More sophisticated papers are those of Franklin (1965a, b).

The first problem one is faced with is how to construct $W(t)$. Many microcomputers and minicomputers have the capability of generating uniformly distributed random numbers on the interval [0, 1]. One then faces the problem of generating a normally distributed random number. This problem is worth addressing since small computers are so common.

8.3. SOLVING SDEs NUMERICALLY

Since $W(t)$ is normally distributed with mean 0 and variance t, $W(t)$ can be represented by $\sqrt{t}\hat{W}$, where $\hat{W}$ is a normal $(0, 1)$ random variable. Suppose that $\{U_i\}$ are the uniform random variables. One way to construct $\hat{W}$ is by an inversion formula. Abramowitz and Stegun (1964, p. 933) give two sets of approximations for $\hat{W}$ such that

$$\frac{1}{\sqrt{2\pi}} \int_{\hat{W}}^{\infty} e^{-s^2/2}\, ds = U. \tag{8.3.2}$$

These approximations are valid for $0 < U \le 0.5$. The first approximation is

$$\hat{W} = l - \left\{\frac{a_0 + a_1 l}{1 + b_1 l + b_2 l^2}\right\} + \varepsilon(U), \tag{8.3.3}$$

where

$$l = [\ln(1/U^2)]^{1/2}, \quad a_0 = 2.30753, \quad a_1 = 0.27061,$$
$$b_1 = 0.99229, \quad b_2 = 0.04481, \quad |\varepsilon(U)| < 0.003. \tag{8.3.4}$$

The second approximation is

$$\hat{W} = l - \left\{\frac{c_0 + c_1 l + c_2 l^2}{1 + d_1 l + d_2 l^2 + d_3 l^3}\right\} + \varepsilon(U), \tag{8.3.5}$$

where

$$l = [\ln(1/U^2)]^{1/2}, \quad |\varepsilon(U)| < 0.00045,$$
$$c_0 = 2.515517, \quad c_2 = 0.802853, \quad c_3 = 0.010328, \tag{8.3.6}$$
$$d_1 = 1.432788, \quad d_2 = 0.189269, \quad d_3 = 0.001308.$$

Exercise 8.3.1

What should be done if $U > 0.5$? How does one construct $W(t)$ from (8.3.3) or (8.3.5)?

The other method for computing $\hat{W}$ is to use the central limit theorem. For example, set

$$\hat{W}_n = \left\{\sum_{i=1}^{n} U_i - \frac{n}{2}\right\}\left\{\frac{n}{12}\right\}^{-1/2}, \tag{8.3.7}$$

so that $\hat{W}_n$ is asymptotically normal $(0, 1)$. When $n = 12$, the maximum errors in $\hat{W}_n$ are 9×10^{-3} for $|\hat{W}_n| < 2$, and 9×10^{-1} for $2 < |\hat{W}_n| < 3$.

Improvements in (8.3.7) are made by using combinations of $\hat{W}_n$. Abramowitz and Stegun (1964, p. 953) suggest

$$\hat{W}_n^* = \hat{W}_n \sum_{s=0}^{k} a_{2s}(\hat{W}_n)^{2s}, \qquad (8.3.8)$$

where, for $n = 12$, the coefficients are

$$\begin{aligned} a_0 &= 9.8746, & a_2 &= 3.9439 \times 10^{-3}, & a_4 &= 7.474 \times 10^{-5}, \\ a_6 &= -5.102 \times 10^{-7}, & a_8 &= 1.41 \times 10^{-7}. \end{aligned} \qquad (8.3.9)$$

In this case the maximum error in the random variable is 8×10^{-4}.

Each of these methods provides a way of approximating $W(t)$. Clearly, the methods give normally distributed random variables with mean 0 and variance t. The only question that remains is the assumption of independence of increments.

Exercise 8.3.2

Derive a procedure that could be numerically implemented to check the assumption of independence of increments.

We shall now solve the SDE (8.3.1). Perhaps the simplest approach to solving (8.3.1) is to rewrite it as a forward difference equation of the form

$$dX = X(t + dt) - X(t) = b(X(t))\, dt + \sqrt{a(X(t))}\,[W(t + dt) - W(t)].$$

Rewriting this equation provides the iteration scheme

$$X(t + dt) = X(t) + b(X(t))\, dt + \sqrt{a(X(t))}\, dW(t),$$

where $dW(t) = W(t + dt) - W(t)$. For many problems this simple iteration scheme will suffice. Franklin (1965a, b) discusses error analyses for equations similar to this one. In many cases the accuracy with this simple scheme is not sufficient and one is interested in more sophisticated ones. Helfand (1979), Rao, Borwankar, and Ramkrishna (1974), and Anderssen, de Hoog, and Weiss (1973) describe various other iteration schemes. It is useful to see how the simplest of these work. Helfand (1979) considers the case in which $b(x, t) = b(x)$ and $a(x, t) = a$ (constant) only. It is a good starting point to review some theory for the deterministic equation

$$dx/dt = b(x), \qquad x(0) = x_0. \qquad (8.3.10)$$

By rewriting (8.3.10) as an integral and Taylor expanding, one obtains

$$\begin{aligned} x(t) = x_0 &+ tb(x_0) + \tfrac{1}{2}t^2 b(x_0) b'(x_0) \\ &+ \tfrac{1}{6}t^3 [b(x_0) b'(x_0)^2 + b(x_0)^2 b''(x_0)] + O(t^4). \end{aligned} \qquad (8.3.11)$$

8.3. SOLVING SDEs NUMERICALLY

Exercise 8.3.3

Derive (8.3.11).

In the deterministic Runge–Kutta (RK) theory, one seeks a solution for (8.3.10) in the form

$$x(t) = x_0 + t(A_1 g_1 + A_2 g_2 + \cdots + A_l g_l). \quad (8.3.12)$$

Here the g_i are motivated by the expansion (8.3.11). In particular, they are given by

$$g_1 = b(x_0),$$
$$g_2 = b(x_0 + \beta_{21} g_1 t),$$
$$g_3 = b(x_0 + \beta_{31} t g_1 + \beta_{32} g_2 t), \quad (8.3.13)$$
$$\vdots$$
$$g_l = b(x_0 + \beta_{l1} t g_1 + \cdots + \beta_{l,l-1} t g_{l-1}).$$

The advantage of an iteration scheme such as (8.3.13) is that one avoids the problem of having to compute derivatives. The parameters β_{ij} and A_k are picked so that the kth-order expansion of (8.3.12) matches (8.3.11). For $k = 2$ the matching gives

$$A_1 + A_2 = 1, \quad A_2 \beta_{21} = 1/2, \quad (8.3.14)$$

so that there is one free parameter.

Exercise 8.3.4

What would be a good criterion to use when picking the adjustable parameter in (8.3.14)?

For the SDE

$$dX = b(X)\,dt + dW, \quad x(0) = x_0, \quad (8.3.15)$$

one tries to mimic the deterministic theory as much as possible. Hence, the first step is to rewrite (8.3.15) as a set of nested integrals given by

$$X(t) = x_0 + \int_0^t dt_1 \, b\!\left(x_0 + \int_0^{t_1} dt_2 \, b(x_0 + \cdots) + W(t_1)\right) + W(t). \quad (8.3.16)$$

The second step involves the Taylor expansion of (8.3.16) up to order t^4. This gives

$$X(t) = x_0 + t b(x_0) + \tfrac{1}{2} t^2 b(x_0) b'(x_0)$$
$$+ \tfrac{1}{6} t^3 [b(x_0) b'(x_0)^2 + b(x_0)^2 b''(x_0)] + \tilde{R} + O(t^4). \quad (8.3.17)$$

In (8.3.17) $\tilde{R}$ is a random term given by

$$\begin{aligned}\tilde{R} = {}& W(t) + b'(x_0)W_1(t) + \frac{1}{2}b''(x_0)\int_0^t W(t_1)^2\, dt \\
& + \Big\{[b'(x_0)]^2 W_2(t) + b(x_0)b''(x_0)[tW_1(t) - W_2(t)] \\
& + \frac{1}{6}b'''(x_0)\int_0^t W(t_1)^3\, dt_1\Big\} \\
& + \Big\{\frac{1}{2}b'(x_0)b''(x_0)\int_0^t dt_1\,(t-t_1)W(t_1)^2 \\
& + \frac{1}{2}b(x_0)b'''(x_0)\int_0^t dt_1\,t_1 W^2(t_1) + \frac{1}{24}b^{(\mathrm{iv})}(x_0)\int_0^t dt_1\, W(t_1)^4\Big\},\end{aligned}$$ (8.3.18)

where

$$W_n(t) = \int_0^t dt_1\, W_{n-1}(t_1), \qquad (8.3.19)$$

with $W_0(t) = W(t)$ (standard Brownian motion).

Exercise 8.3.5

Verify (8.3.17) by Taylor expanding (8.3.16). In addition, show that the terms in braces in (8.3.18) are all the same order in t.

The functions $W_n(t)$ are iterates of Brownian motion. In analogy with the deterministic theory, one tries to match the expansion (8.3.17) with a guessed but simpler expansion. To do this, moments of $\tilde{R}$ are needed. In order to find them, the moments of $W_n(t)$ are needed.

Exercise 8.3.6

Show that

$$E\{W_n(t)W_m(t)\} = t^{n+m+1}/n!m!(n+m+1), \qquad n, m \geq 1;$$

$$E\{W_0(t')W_1(t)\} = \begin{cases}\frac{1}{2}t^2, & t \leq t', \\ \frac{1}{2}t'(2t - t'), & t \geq t'.\end{cases} \qquad (8.3.20)$$

The moments of the random term $\tilde{R}$ are easily found. They are given by

$$\begin{aligned}E\{R\} &= \tfrac{1}{4}t^2 b''(x_0) + t^3[\tfrac{1}{12}b'(x_0)b''(x_0) + \tfrac{1}{6}b(x_0)b'''(x_0) + \tfrac{1}{24}b^{(\mathrm{iv})}(x_0)] + O(t^4), \\
E\{R^2\} &= t + t^2 b'(x_0) + t^3[\tfrac{2}{3}b'(x_0)^2 + \tfrac{2}{3}b(x_0)b''(x_0) + \tfrac{1}{3}b'''(x_0)] + O(t^4), \\
E\{R^3\} &= \tfrac{7}{4}t^3 b''(x_0) + O(t^4).\end{aligned} \qquad (8.3.21)$$

8.3. SOLVING SDEs NUMERICALLY

The second-order Runge–Kutta scheme assumes that $X(t)$ can be represented by the form

$$X(t) = x_0 + t(A_1 g_1 + A_2 g_2) + \sqrt{t} \lambda_0 \tilde{Z}, \qquad (8.3.22)$$

where $\tilde{Z} \sim N(0, 1)$ and the $g_i(x)$ are given by

$$\begin{aligned} g_1 &= b(x_0 + \sqrt{t} \lambda_1 \tilde{Z}), \\ g_2 &= b(x_0 + t\beta g_1 + \sqrt{t} \lambda_2 \tilde{Z}). \end{aligned} \qquad (8.3.23)$$

One picks the parameters by matching solutions and moments. That is, if (8.3.22) is Taylor expanded, then

$$\begin{aligned} X(t) = {}& x_0 + (A_1 + A_2)tb(x_0) + A_2 \beta t^2 b(x_0) b''(x_0) \\ & + \lambda_0 \tilde{Z}\sqrt{t} + (A_1 \lambda_1 + A_2 \lambda_2) \tilde{Z} t^{3/2} b'(x_0) \\ & + \tfrac{1}{2}(A_1 \lambda_1^2 + A_2 \lambda_2^2) \tilde{Z}^2 t^2 b''(x_0) + O(t^3). \end{aligned} \qquad (8.3.24)$$

Thus, when the deterministic part of (8.3.24) is matched to that of (8.3.17) and the first two central moments of (8.3.24) are matched to the first two central moments in (8.3.21), one obtains

$$\begin{gathered} A_1 + A_2 = 1, \qquad A_2 \beta = \tfrac{1}{2}, \qquad \lambda_0^2 = 1, \\ (A_1 \lambda_1 + A_2 \lambda_2)\lambda_0 = \tfrac{1}{2}, \qquad A_1 \lambda_1^2 + A_2 \lambda_2^2 = \tfrac{1}{2}. \end{gathered} \qquad (8.3.25)$$

There are six parameters and five equations, so one parameter is adjustable. Helfand (1979) suggests the choice of parameters

$$\begin{gathered} A_1 = A_2 = \tfrac{1}{2}, \qquad \beta = 1 = \lambda_0, \\ \lambda_1 = 0, \qquad \lambda_2 = 1, \quad \text{or} \quad \lambda_1 = 1, \qquad \lambda_2 = 0. \end{gathered} \qquad (8.3.26)$$

Helfand (1979) also gives a third-order scheme and indicates how to find higher-order schemes.

Bibliographic Notes

General introductions to the numerical solution of differential equations are found in Burden, Faires, and Reynolds (1978), Blum (1972), and Ortega (1972). The algorithms discussed in this section are generalized to include state-dependent diffusion coefficients by Rao, Borwankar, and Ramkrishna (1974) and Anderssen, de Hoog, and Weiss (1973). Franklin's papers (1965a, b) are very nice introductions to numerical simulation for stochastic differential equations.

8.4 SOME NUMERICAL CONSIDERATIONS FOR DPEs

In this final section of the final chapter, numerical solutions of DPEs are considered. The caveat that introduced this chapter is repeated: each problem has its own details and difficulties. This section can only provide general guidelines and a background in some of the technical detail.

The time-dependent DPE considered in this section takes the form

$$0 = J_t + \max_u \{b(x, t, u)J_x + e^{-\rho t}L(x, t, u)\} + \tfrac{1}{2}a(x, t)J_{xx}$$

$$x \geq 0, \quad t \geq 0. \quad (8.4.1)$$

The time-independent DPE takes the form

$$0 = -\rho J + \max_u \{b(x, u)J_x + L(x, u)\} + \tfrac{1}{2}a(x)J_{xx}, \quad x \geq 0. \quad (8.4.2)$$

Since there is no general theory, particular cases will be studied.

First, note that sometimes the form of $b(\cdot)$ and $a(\cdot)$ is such that computational difficulties occur near $x = 0$. Ludwig (1980) suggests the change of variables $y = \ln x$. Then, for example, (8.4.1) becomes

$$0 = J_t + \max_u \left\{ \frac{b(e^y, t, u)}{e^y} J_y + e^{-\rho t}L(e^y, t, u) \right\}$$

$$+ \frac{a(e^y, t)}{2} e^{-2y} J_{yy} - e^{-2y} \frac{a(e^y, t)}{2} J_y. \quad (8.4.3)$$

Ludwig suggests imposing boundary conditions on y of the form

$$\left[\frac{b(e^y, t, u)}{e^y} - e^{2y} \frac{a(e^y, t)}{2} \right] J_y + e^{-\rho t}L(e^y, t, u) = 0 \quad (8.4.4)$$

at $y = y_1, y_2$ ($y_1 < 0 < y_2$). This corresponds to

$$b(x, t, u)J_x + e^{-\rho t}L(x, t, u) = 0 \quad (8.4.5)$$

at $x = e^{y_1}, e^{y_2}$ for the original equation. These boundary conditions essentially ignore fluctuations for large and small values of x.

Exercise 8.4.1

When would boundary conditions such as (8.4.4) or (8.4.5) be justified?

One way to solve DPEs is by discretizing them in the domain of interest. In general, such procedures are very difficult to implement. One successful example is presented by van Mellaert and Dorato (1972), who attribute the

8.4. SOME NUMERICAL CONSIDERATIONS FOR DPEs

method to Samarskii (1966). The DPE considered takes the very special form

$$v_t = \sum_{i=1}^{m} \tfrac{1}{2} a_i v_{x_i x_i} + b_i v_{x_i} + A \operatorname{sgn}(v_{x_m}) v_{x_m}, \quad x = (x_1, x_2, \ldots, x_m); \quad (8.4.6)$$

$$v(x, t) = 0, \quad x \in \text{boundary},$$
$$v(x, 0) = 1, \quad x \in \text{the domain } D. \quad (8.4.7)$$

Here sgn(x) is the sign (+ or −) of x.

Exercise 8.4.2

Equation (8.4.6) corresponds to the DPE for a highly specialized stochastic control problem. What is this problem and the nature of the control?

The procedure of van Mellaert and Dorato splits (8.4.6) into a set of m equations:

$$\frac{1}{m} v_t = \frac{a_i}{2} v_{x_i x_i} + [b_i(x) + \delta_{im} u_0] v_{x_i}, \quad i = 1, 2, \ldots, m, \quad (8.4.8)$$

where δ_{im} is the Kroenecker delta function and

$$u_0 = A \operatorname{sgn}(v_{x_m}). \quad (8.4.9)$$

Each of the equations (8.4.8) is solved along a line according to the following rules. Span D by an m-dimensional set of lines spaced h units apart. Discretize time in units of τ.

(1) For x_1 discretize (8.4.8) to

$$\frac{v(\tau/m) - v(0)}{\tau} = \frac{a_1}{2} \left(\frac{v_1^+(\tau/m) - 2v(\tau/m) + v_1^-(\tau, m)}{h^2} \right)$$
$$+ \frac{b_1(v_1^+(\tau/m) - v_1^-(\tau/m))}{2h}. \quad (8.4.10)$$

Here the following notation is introduced

$$v(\tau/m) = v(x_1, \ldots, x_m, \tau/m),$$
$$v_1^+(\tau/m) = v(x_1 + h, \ldots, x_m, \tau/m), \quad (8.4.11)$$
$$v_1^-(\tau/m) = v(x_1 - h, \ldots, x_m, \tau/m).$$

The boundary condition (8.4.7) is matched by linear extrapolation.

(2) For x_2 discretize (8.4.8) to

$$\frac{v(2\tau/m) - v(\tau/m)}{\tau} = \frac{a_2}{2}\left[\frac{v_2^+(2\tau/m) - 2v(2\tau/m) + v_2^-(2\tau/m)}{h^2}\right]$$
$$+ b_2\left[\frac{v_2^+(2\tau/m) - v_2^-(2\tau/m)}{2h}\right], \qquad (8.4.12)$$

where $v(\tau/m)$ is obtained from step 1.

(3) Continue this procedure until x_m is reached. For x_m the discretized version is

$$\frac{v(\tau) - v(\tau(m-1)/m)}{\tau} = \frac{a_m}{2}\left[\frac{v_m^+(\tau) - 2v(\tau) + v_m^-(\tau)}{h^2}\right]$$
$$+ \left[b_m + A \, \text{sgn}\left\{v_m^+\left(\frac{m-1}{m}\tau\right) - v_m^-\left(\frac{m-1}{m}\tau\right)\right\}\right]$$
$$\times \left[\frac{v_m^+(\tau) - v_m^-(\tau)}{2h}\right]. \qquad (8.4.13)$$

(4) One trip through steps 1–3 gives an approximation for $v(x, \tau)$. These steps are repeated for $t = 2\tau, 3\tau, \ldots$.

The advantage of this method is that the nonlinear control term in (8.4.13) is known from the previous step. Thus, all of the equations are linear. A distinct disadvantage of this method is the special structure that it requires of the problem. In particular, the special way in which the control enters into the formulation affects the general usefulness of the solution.

Another method for solution of DPEs is that of *splines* [see, e.g., Blum (1972) or Burden, Faires, and Reynolds (1978)]. In this section cubic splines are discussed.

As an example, consider the DPE

$$\tfrac{1}{2}a(x)J_{xx} + \max_u\{b(x, u)J_x + L(x, u)\} - \rho J = 0 \qquad (8.4.14)$$

on $0 < l_1 \le x \le l_2$. Discretize the interval $[l_1, l_2]$ into N pieces and let $\{x_k\}$ denote a set of mesh points. For example, for a uniform mesh

$$x_k = l_1 + \frac{k(l_2 - l_1)}{N}, \qquad k = 0, 1, 2, \ldots, N. \qquad (8.4.15)$$

The basic idea here is to assume that on $[x_k, x_{k+1}]$, $J(x)$ can be approximated by a function $J_k(x)$ which is defined by

$$J_k(x) = \alpha_k + \beta_k(x - x_k) + \gamma_k(x - x_k)^2 + \delta_k(x - x_k)^3. \qquad (8.4.16)$$

8.4. SOME NUMERICAL CONSIDERATIONS FOR DPEs

The task at hand is to determine the set of coefficients $\{\alpha_k, \beta_k, \gamma_k, \delta_k\}$, $k = 0, 1, \ldots, N$. Since the solution of (8.4.14) should be continuous and differentiable for interior x_k, it must be that the following conditions hold:

$$J_{k+1}(x_{k+1}) = J_k(x_{k+1}),$$

$$\left.\frac{\partial J_{k+1}}{\partial x}\right|_{x_{k+1}} = \left.\frac{\partial J_k}{\partial x}\right|_{x_{k+1}}, \qquad (8.4.17)$$

$$\left.\frac{\partial^2 J_{k+1}}{\partial x^2}\right|_{x_{k+1}} = \left.\frac{\partial^2 J_k}{\partial x^2}\right|_{x_{k+1}}.$$

At x_0, x_N the spline J_0 or J_N must match the boundary condition. The three conditions (8.4.17) give (with $\Delta_k \equiv x_{k+1} - x_k$) three conditions on the coefficients for interior points. These conditions are

$$\alpha_{k+1} = \alpha_k + \beta_k \Delta_k + \gamma_k \Delta_k^2 + \delta_k \Delta_k^3, \qquad (8.4.18)$$

$$\beta_{k+1} = \beta_k + 2\gamma_k \Delta_k + 3\delta_k \Delta_k^2, \qquad (8.4.19)$$

$$\gamma_{k+1} = \gamma_k + 3\delta_k \Delta_k. \qquad (8.4.20)$$

The fourth condition on the coefficients is determined from the DPE itself, since J_{k+1} and J_k both satisfy the DPE (8.4.14). This gives

$$a(x_{k+1})\gamma_{k+1} + \max_u\{b(x_{k+1})\beta_{k+1} + L(x_{k+1}, u)\} - \rho \alpha_{k+1}$$

$$= \tfrac{1}{2}a(x_{k+1})\{2\gamma_k + 6\delta_k \Delta_k\}$$

$$\quad + \max_u\{b(x_{k+1})(\beta_k + 2\gamma_k \Delta_k + 3\delta_k \Delta_k^2) + L(x_{k+1}, u)\}$$

$$\quad - \rho(\alpha_k + \beta_k \Delta_k + \gamma_k \Delta_k^2 + \delta_k \Delta_k^3) = 0. \qquad (8.4.21)$$

Exercise 8.4.3

What requirements on the coefficients are obtained from the condition $J_k(x_k) = J_{k-1}(x_k)$.

Solving (8.4.20) for δ_k gives

$$\delta_k = (\gamma_{k+1} - \gamma_k)/3\Delta_k. \qquad (8.4.22)$$

Using (8.4.22) in (8.4.19) allows one to get an equation for γ_k. It is

$$\gamma_k = (\beta_{k+1} - \beta_k - \Delta_k \gamma_{k+1})/\Delta_k. \qquad (8.4.23)$$

Exercise 8.4.4

Substitute (8.4.23) into (8.4.18) to obtain the equation for β_k; then use (8.4.21) to obtain an equation for α_k.

The set of recursive equations similar to (8.4.22, 23) allows one to determine the coefficients. In practice, such computations require considerable skill and imagination.

Bibliographic Notes

Ludwig (1979b) discusses how the method of splines can be used to solve stochastic control problems. The splines introduced here are cubic splines, which are discussed in the books by Blum (1972), Burden, Faires, and Reynolds (1978), and Stoer and Bulirsch (1980). Quintic splines have also been used in some stochastic control problems. Banks (1980) and Kalaba and Springarn (1982) provide an introduction to such methods. Other approaches for the numerical solution of DPEs are discussed by Mendelssohn (1978), Norman and White (1968), Szidarovsky and Yakowitz (1978), and Yakowitz (1969).

References

Abramowitz, M., and Stegun, I. (1964). "Handbook of Mathematical Functions," National Bureau of Standards, Washington, D. C.

Agnello, R. J., and Anderson, L. G. (1981). Production responses for multispecies fisheries. *Can. J. Fish. Aqua Sci.* **38**, 1393–1404.

Allais, M (1957). Method of appraising economic prospects of mining exploration over large territories. *Management Sci.* **3**, 285–347.

Anderssen, R. S., de Hoog, F. R., and Weiss, R. (1973). On the numerical solution of Brownian motion processes. *J. Appl. Probab.* **10**, 409–418.

Anderson, D. R. (1975).Optimal exploitation strategies for an animal population in a Markovian environment: a theory and an example. *Ecology* **56**, 1281–1297.

Anderson, J. R., Dillon, J. L., and Hardaker, B. (1977). "Agricultural Decision Analysis." Iowa State Univ. Press, Ames, Iowa.

Anderson, R. M., Turner, B. D., and Taylor L. R., eds. (1979). "Population Dynamics." Blackwell, London.

Aoki, M. (1967). "Optimization of Stochastic Systems." Academic Press, New York.

Aris, R., and Amundson, N. R. (1973). "Mathematical Methods in Chemical Engineering," vol 2. Prentice-Hall, Englewood Cliffs, New Jersey.

Arkin, V. I., Kolemaev, V. A., and Sirjaev, A. N. (1966). On finding optimal controls. *Select. Transl. Math. Stat. Prob.* **6**, 55–60.

Arnold, L. (1974). "Stochastic Differential Equations." Wiley, New York.

Arnold, V. I. (1972). Lectures on bifurcations in versal families. *Russian Math. Surveys* **27**, 54–123.

Arrow, K. J. (1962), The economic implications of learning by doing. *Rev. Econom. Stud.* **29**, 155–173.

Arrow, K. J., and Chang, S. S. L. (1980). Optimal pricing, use, and exploration of uncertain natural resource stocks. *In* "Dynamic Optimization and Mathematical Economics" (P. T. Liu, ed.), pp. 105–116. Plenum, New York. Also see *J. Environ. Econom. Man.* **9**, 1–10 (1982).

Athans, M. (1972). On the determination of optimal costly measurement strategies for linear stochastic systems. *Automatica—J. IFAC* **8**, 397–412.

Avriel, M. (1976). "Nonlinear Programming: Analysis and Methods." Prentice-Hall, Englewood Cliffs, New Jersey.

Banks, H. T. (1980). Computational difficulties in the identification and optimization of control systems, *In* "Renewable Resource Management" (T. L. Vincent and J. M. Skowronski, eds.), pp. 79–94. Springer-Verlag, Berlin and New York.

Baram, Y. and Sandell, N. R. (1978). An information theoretic approach to dynamical systems modeling and identification. *IEEE Trans. Automat. Control* **AC-23**, 61–66.
Barkley, R. A. (1964). The theoretical effectiveness of towed net samplers as related to sample size and to swimming speed of organisms. *J. Cons. Int. Explor. Mer.* **29**, 146–157.
Barlow, R. E., Bartholomew, D. J., Bremner, J. M., and Brunk, H. D. (1972). "Statistical Inference Under Order Restrictions." Wiley, New York.
Bar Shalom, Y., and Tse, E. (1976). Concepts and methods in stochastic control. *In* "Control and Dynamic Systems." (C. T. Leondes, ed.), pp. 99–172, Vol. 12, Academic Press, New York.
Bartlett, B. R. (1968). Outbreaks of two spotted spider-mites and cotton aphids following pesticide treatment. I. Pest stimulation vs. natural enemy destruction as the cause of outbreaks. *J. Econ. Entom.* **61**, 297–303.
Bartlett, M. S. (1964). The spectral analysis of two dimensional point processes. *Biometrika* **51**, 299–311.
Bartlett, M. S. (1974). The statistical analysis of spatial pattern. *Adv. Appl. Probab.* **6**, 336–358.
Bather, J. A. (1981). Randomized allocation of treatments in sequential experiments. *J. Roy. Statist. Soc. Ser. B* **43**, 265–292.
Beddington, J. R., and Clark, C. W. (1984). Allocation problems between national and foreign fisheries. *Mar. Res. Econ.* **1**, forthcoming.
Bell, D. E., ed. (1977). "Conflicting Objectives in Decisions." Wiley, New York.
Bellman, R. (1957). "Dynamic Programming." Princeton Univ. Press, Princeton, New Jersey.
Bellman, R., and Dreyfus, S. (1962). "Applied Dynamic Programming." Princeton Univ. Press, Princeton, New Jersey.
Bender, C. M., and Orszag S. A. (1978). "Advanced Mathematical Methods for Scientists and Engineers." McGraw-Hill, New York.
Berger, J. O. (1980), "Statistical Decision Theory." Springer-Verlag, Berlin and New York.
Bertsekas, D. (1976). "Dynamic Programming and Stochastic Control." Academic Press, New York.
Best, P. B., and Butterworth, D. S. (1980). Report of the southern hemisphere Minke whale assessment cruise, 1978/79. *Rep. Int. Whale Comm.* **30**, 257–283.
Beverton, R. J. H., and Holt, S. J. (1957). "On the Dynamics of Exploited Fish Populations." Ministry of Agriculture, Fisheries and Food (London), Fisheries Investigation Series 2(19).
Blum, E. K. (1972). "Numerical Analysis and Computation Theory and Practice." Addison-Wesley, Reading, Massachusetts.
Boyce, W. E., and DiPrima, R. C. (1977). "Elementary Differential Equations and Boundary Value Problems." Wiley, New York.
Bronson, R. (1982). "Operations Research." McGraw-Hill, New York.
Bryson, A. E. and Ho, Y. C. (1975). "Applied Optimal Control." Halsted Press, New York.
Burden, R. L., Faires, J. D., and Reynolds, A. C. (1978). "Numerical Analysis." Prindle Weber and Schmidt, Boston, Massachusetts.
Burnham, K. P., Anderson, D. R., and Laake, J. L. (1980). "Estimation of density from line transect sampling of biological populations. *Wildlife Monogr.* **72**, 202 pp.
Butterworth, D. S. (1982). A possible basis for choosing a functional form for the distribution of sightings with right angle distance: some preliminary ideas. *Rep. Int. Whale Comm.* **32**, 555–558.
Carey, J. R. (1982a). Within plant distribution of tetranychid mites of cotton. *Environ. Entomol.* **11**, 796–800.
Carey, J. R. (1982b). Demography of the two spotted spider mite Tetranychus urticae Koch. *Oecologia* **52**, 389–395.

REFERENCES

Carey, J. R., and Bradley, J. W. (1982). Developmental rates, vital schedules, sex ratios and life tables for *Tetranychus Urticae*, *T. Turkestani*, and *T. Pacificus* (*Acarina: Tetranychidae*) on cotton, *Acarologia* **23**, 333–345.

Charlesworth, B. (1980). "Evolution in Age Structured Populations." Cambridge Univ. Press, London and New York.

Charnes, A., and Cooper, W. W. (1958). The theory of search: optimum distribution of search effort, *Management Sci.* **5**, 44–50.

Chernoff, H., and Petkau, A. J. (1978). Optimal control of a Brownian motion. *SIAM J. Appl. Math.* **34**, 717–731.

Chow, G. C. (1981). "Econometric Analysis by Control Methods." Wiley, New York.

Clark, C. W. (1976). "Mathematical Bioeconomics." Wiley, New York.

Clark, C. W., and Mangel, M. (1979). Aggregation and fishery dynamics: a theoretical study of schooling and the purse-seine fishery. *Fish. Bull.* **77**, 317–337.

Clark, C. W., Charles, A., Beddington, J. R., and Mangel, M. (1984). Optimal capacity decisions in a developing fishery. *Marine Res. Econ.* **1**, forthcoming.

Cochran, W. G. (1977). "Sampling Techniques." Wiley, New York.

Coddington, E., and Levinson, N. (1955). "Theory of Ordinary Differential Equations." McGraw-Hill, New York.

Comins, H. N. (1977a). The development of insecticide resistance in the presence of migration. *J. Theoret. Biol.* **64**, 177–197.

Comins, H. N. (1977b). The management of pesticide resistance. *J. Theoret. Biol.* **65**, 399–420.

Conklin, F. S., Baquet, A. E., and Halter, A. N. (1977). "A Bayesian Simulation Approach for Estimating the Value of Information: An Application to Frost Forecasting." Technical Bulletin 136, Agricultural Experiment Station. Oregon State University, Corvallis, Oregon.

Cooke, F. T., and Parvin, D. W. (1981). "Insecticide Use of Cotton in the United States—1969, 1972, and 1974." ERS Staff Report No. AGESS–810331, U. S. Department of Agriculture, Economics Research Section, Natural Resource Economics Division, Washington, D. C.

Courant, R., and Hilbert, D. (1953). "Methods of Mathematical Physics," vol. 1. Wiley, New York.

Courant, R., and Hilbert, D. (1962). "Methods of Mathematical Physics," vol. 2. Wiley, New York.

Cozzolino, J. M. (1970). Sequential search for an unknown number of objects of non-uniform size. *Oper. Res.* **20**, 293–308.

Cozzolino, J. M. (1977). A new method for measurement and control of exploration risk. *Soc. Pet. Eng. J.* **6632**, 1–8.

Cozzolino, J. M. (1979). Measurement and projection of search efficiency. *Soc. Pet. Eng. J.* **7456**, 251–256.

Cozzolino, J. M., and Falconer, W. A. (1977). Williston basin search area analyzed. *Oil Gas J.*, 17 January.

Crawford, R. G. (1973). Implications of learning for economic models of uncertainty. *Internat. Econom. Rev.* **14**, 587–600.

Crow, J., and Kimura, M. (1979). "An Introduction to Population Genetics Theory." Harper and Row, New York.

Cushing, D. H. (1981). "Fisheries Biology." University of Wisconsin Press, Madison, Wisconsin.

Dasgupta, P. S., and Heal, G. M. (1979). "Economic Theory and Exhaustible Resources." Nisbet-Cambridge, Digswell Place, Welwyn, Herts, United Kindom.

Davis, M. H. A. (1977). "Linear Estimation and Stochastic Control." Chapman and Hall, London.

DeGroot, M. H. (1970). "Optimal Statistical Decisions." McGraw-Hill, New York.
DeGroot, M. H. (1975). "Probability and Statistics." Addison-Wesley, Reading, Massachusetts.
Decker, D. W., and Kelley, C. T. (1980). Newton's method at singular points. *SIAM J. Numer. Anal.* **17**, 66–70.
Derzko, N. A., and Sethi, S. P. (1981a). Optimal exploration and consumption of a natural resource stock—stochastic case. *Int. J. Policy Anal. Info. Sys.* **5**, 185–200.
Derzko, N. A., and Sethi, S. P. (1981b). Optimal exploration and consumption of a natural resource—deterministic case. *Optimal Control Appl. Methods* **2**, 1–21.
Devarajan, S., and Fisher, A. C. (1981). Hotelling's "Economics of Exhaustible Resources": fifty years later. *J. Econom. Lit.* **190**, 65–73.
Diamond, P., and Rothschild, M. (1978). "Uncertainty in Economics." Academic Press, New York.
Dixon, B. L., and Howitt, R. E. (1979a). Continuous forest inventory using a linear filter. *For. Sci.* **25**, 675–689.
Dixon, B. L., and Howitt, R. E. (1979b). "Uncertainty and the Intertemporal Management of Natural Resources: An Empirical Application to the Stanislaus National Forest." Giannini Foundation Monograph 38, University of California, Division of Agricultural Sciences, Berkeley, California.
Dorfman, R. (1969). An economic interpretation of optimal control theory. *Amer. Econom. Rev.* **59**, 817–831.
Draper, N. R., and Smith, H. (1981). "Applied Regression Analysis." Wiley, New York.
Dreyfus, S. (1972). The main results of optimal control theory made simple, *In* "Population Dynamics" (T. N. E. Grenville, ed.). Academic Press, New York.
Dreyfus, S., and Law, A. (1977). "The Art and Theory of Dynamic Programming." Academic Press, New York.
Duda, R. O., and Hart, P. E. (1973) "Pattern Classification and Scene Analysis." Wiley, New York.
Duffie, D., and Taskar, M. (1983). Diffusion approximation in Arrow's model of exhaustible resources. Technical Report 416, Institute for Mathematical Studies in the Social Sciences, Stanford University, Stanford, California.
Eggers, D. M. (1979). "Design of Eastern Bering Sea Trawl Survey: Allocation of Effort among Strata." International North Pacific Commission Document No. 2241, Seattle, Washington.
Englbrecht-Wiggans, R. (1980). Auctions and bidding models: a survey. *Management Sci.* **26**, 119–142.
Epple, D., and Lave, L. B. (1982). The Helium storage controversy: modeling natural resource supply. *Amer. Sci.* **70**, 286–293.
Everhart, W. H., and Youngs, W. D. (1981). "Principles of Fisheries Science." Cornell Univ. Press, Ithaca, New York.
Feller, W. (1968). "An Introduction to Probability Theory and Its Applications," vol. 1. Wiley, New York.
Feller, W. (1971). "An Introduction to Probability Theory and Its Applications," vol. 2. Wiley, New York.
Fleming, W., and Rishel, R. W. (1975). "Deterministic and Stochastic Optimal Control." Springer-Verlag, Berlin and New York.
Foerster, R. E., and Ricker, W. E. (1941). The effect of reduction of predaceous fish on survival of young sockeye salmon at Cultus Lake. *J. Fish. Res. Bd. Can.* **5**, 315–336.
Franklin, J. N. (1965a). Difference methods for stochastic ordinary differential equations. *Math. Comp.* **19**, 552–561.

REFERENCES

Franklin, J. N. (1965b). Numerical simulation of stationary and nonstationary Gaussian random processes. *SIAM Rev.* **7**, 68–80.
Gelb, A., ed. (1974). "Applied Optimal Estimation." MIT Press, Cambridge, Massachusetts.
Georghiou, G. P. (1980). Insecticide resistance and prospects for its management. *Residue Rev.* **76**, 131–145.
Georghiou, G. P., and Taylor, C. E. (1977a). Operational influences in the evolution of insecticide resistance. *J. Econom. Entomol.* **70**, 653–658.
Georghiou, G. P., and Taylor, C. E. (1977b). Genetic and biological influences in the evolution of insecticide resistance. *J. Econom. Entomol.* **70**, 319–325.
Gilbert, R. J. (1976). Search strategies for nonrenewable resource deposits. Technical Report 196, Institute for Mathematical Studies in the Social Sciences, Stanford University, Stanford, California.
Gilbert, R. J. (1979). Optimal depletion of an uncertain stock. *Rev. Econom. Studies* **46**, 47–57.
Gilbert, R. J., and Richels, R. G. (1981). Reducing uranium resource uncertainty. Is it worth the cost? *Res. Energy* **3**, 13–37.
Gleit, A. (1978). Optimal harvesting in continuous time with stochastic growth. *Math. Biosci.* **41**, 111–123.
Gordon, H. S. (1954). Economic theory of a common property resource: the fishery. *J. Polit. Econom.* **75**, 274–286.
Green, R. F. (1980). Bayesian birds: a simple example of Oaten's model of optimal foraging. *Theoret. Population Biol.* **18**, 244–256.
Griffiths, J. C. (1966). Exploration for natural resources. *Oper. Res.* **14**, 189–209.
Gulland, J. A., ed. (1977). "Fish Population Dynamics." Wiley, New York.
Gutierrez, A. P., Falcon, L. A. Loew, W., Leipzig, P. A., and van den Bosch, R. (1975). An analysis of cotton production in California: model for Acala cotton and the effects of defoliators on its yields. *Env. Entomol.* **4**, 125–136.
Hagan, P. S., Caflisch, R. E., and Keller, J. B. (1981). Arrow's model of optimal pricing, use and exploration of uncertain natural resources. Preprint, Department of Mathematics, Stanford University, Stanford, California.
Haimes, Y. Y. (1977). "Hierarchial Analysis of Water Resource Systems." McGraw-Hill, New York.
Harbaugh, J. W., Doveton, J. H., and Davis, J. C. (1977). "Probability Methods in Oil Exploration." Wiley, New York.
Harris, D. P., and Skinner, B. J. (1982). The assessment of long term supplies of minerals. *In* "Explorations in Natural Resource Economics." (V. K. Smith and J. V. Krutilla, eds.). Johns Hopkins Press, Baltimore, Maryland.
Hartwick, J. M. (1978). Exploitation of many deposits of an exhuastible resource. *Econometrica* **46**, 201–217.
Hassell, M. P. (1978). "The Dynamics of Arthropod Predator–Prey Systems." Princeton Univ. Press, Princeton, New Jersey.
Helfand, E. (1979). Numerical integration of stochastic differential equations. *Bell. System Tech. J.* **58**, 2289–2299.
Helle, W., and van Zon, A. D. (1967). Rates of spontaneous mutation in certain genes of an arrhenotokus mite, *Tetrachynus Pacificus*. *Entomol. Exp. Appl.* **10**, 189–193.
Hestenes, M. R. (1980)."Calculus of Variations and Optimal Control Theory." R. F. Krieger, Huntington, New York.
Heyman, D. P., and Sobel, M. (1984). "Stochastic Models in Operations Research," vol. II. McGraw-Hill, New York.
Hirshleifer, J. (1970). "Investment, Interest, and Capital." Prentice-Hall, Englewood Cliffs, New Jersey.

Hoel, M. (1978a). Resource extraction, uncertainty, and learning. *Bell J. Econom.* **9**, 642–645.
Hoel, M. (1978b). Resource extraction when a future substitute has an uncertain cost. *Rev. Econom. Studies* **45**, 637–644.
Hoel. M. (1979). "Extraction of an Exhaustible Resource Under Uncertainty." Oelgeschlager, Gunn and Hain, Cambridge, Massachusetts.
Hoel, M. (1979). Resource extraction, uncertainty and learning. Memorandum, Institute of Economics, University of Oslo, Oslo, Norway.
Hoel, P. G., Port, S. C., and Stone, C. J. (1971). "Introduction to Probability Theory." Houghton Mifflin, Boston, Massachusetts.
Hoppenstaedt, F. (1975). "Mathematical Theories of Populations: Demographics, Genetics and Epidemics." *Soc. Ind. Appl. Math.*, Philadelphia, Pennsylvania.
Hotelling, H. (1931). The economics of exhaustible resources. *J. Pol. Econom.* **39**, 137–175.
Hueth, D., and Regev, U. (1974). Optimal agricultural pest management with increasing pest resistance. *Amer. J. Ag. Econom.* **56**, 543–552.
Hutchinson, C. E., and Fischer, T. R. (1979). Stochastic control theory applied to fishery management. *IEEE Trans. Systems Man. Cybernet.* **SMC-9**, 253–259.
Intriligator, M. D. (1971). "Mathematical Optimization and Economic Theory." Prentice-Hall, Englewood Cliffs, New Jersey.
Jacobson, D., and Mayne, D. (1970). "Differential Dynamic Programming." Academic Press, New York.
Janis, I., and Mann, L. (1977). "Decision Making." Free Press, MacMillan, New York.
John, F. (1978). "Partial Differential Equations." Springer-Verlag, Berlin and New York.
Judge, G. G., Griffiths, W. E., Hill, R. C., and Lee, T. S. (1980). "The Theory and Practice of Econometrics." Wiley, New York.
Kahneman, D., and Tversky, A. (1979). Prospect theory: an analysis of decision under risk. *Econometrica* **47**, 263–291.
Kalaba, R., and Spingarn, K. (1982). "Control, Identification, and Input Optimization." Plenum, New York.
Kalman, R. E. (1960). A new approach to linear filtering and prediction problems. *Trans. ASME* **82D**, 35–45.
Kamien, M. I., and Schwartz, N. L. (1981). "Dynamic Optimization: The Calculus of Variations and Optimal Control Theory in Economics and Management." North-Holland Publ., Amsterdam.
Kamil, A. C., and Sargent, T. D. (1981). "Foraging Behavior." Garland Press, New York.
Karlin, S., and Taylor, H. M. (1977). "A First Course in Stochastic Processes." Academic Press, New York.
Karlin, S., and Taylor, H. M. (1981). "A Second Course in Stochastic Processes." Academic Press, New York.
Kendall, M., and Stuart A. (1973). "The Advanced Theory of Statistics." vol. 2. Griffin and Company, London.
Kleindorfer, P. R. (1978). Stochastic control models in management science: theory and computation. *TIMS Studies in the Management Sciences* **9**, 69–88.
Knessl, C., Mangel, M., Matkowsky, B. J., Schuss, Z., and Tier, C. (1984). Solution of Kramers–Moyal equations for problems in chemical physics. *J. Chem. Phys.* **81**, 1285–1293.
Knight, F. W. (1971). "Risk, Uncertainty, and Profit." Univ. of Chicago Press, Chicago Illinois.
Kolbin, V. V. (1971). "Stochastic Programming." D. Reidel, Boston, Massachusetts.
Koopman, B. O. (1980). "Search and Screening." Pergamon, Oxford.
Kottegoda, N. T. (1980). "Stochastic Water Resources Technology." Wiley, New York.
Krebs, J. R., and Davies, N. B. (1978). "Behavioural Ecology." Blackwell Scientific Publications, Oxford Univ. Press, London and New York.

REFERENCES

Krebs, J. R., Kacelnik, A., and Taylor, P. (1978). Test of optimal sampling by foraging Great Tits. *Nature* **275**, 27–31.
Larkin, P. A. (1977). Pacific Salmon. *In* "Fish Population Dynamics." (J. A. Gulland, ed.), pp. 156–181. Wiley, New York.
Leaman, B. M. (1981). A brief review of survey methodology with regard to groundfish stock assessment. *Canad. Spec. Publ. Fish. Aq. Sci.* **58**, 113–123.
Leitman, G. (1981). "The Calculus of Variations and Optimal Control." Plenum, New York.
Lewis, T. (1982). Exploitation of a renewable resource under uncertainty. *Canad. J. Econom.* **14**, 422–439.
Lighthill, M. J. (1975). "Introduction to Fourier Analysis and Generalized Functions." Cambridge Univ. Press, London and New York.
Lin C. C., and Segel, L. A. (1974). "Mathematics Applied to Deterministic Problems in the Natural Sciences." MacMillan, New York.
Loucks, D. P., Stedinger, J. R., and Haith, D. A. (1981). "Water Resource Systems Planning and Analysis." Prentice-Hall, Englewood Cliffs, New Jersey.
Loury, G. C. (1978). The optimal exploitation of an unknown reserve. *Rev. Econom. Studies* **45**, 621–636.
Ludwig, D. (1974). "Stochastic Population Theories." Springer-Verlag, Berlin and New York.
Ludwig, D. (1975). Persistence of dynamical systems under random perturbations. *SIAM Rev.* **17**, 605–640.
Ludwig, D. (1979a). Optimal harvesting of a randomly fluctuating resource. I: Application of perturbation methods. *SIAM J. Appl. Math.* **37**, 166–184.
Ludwig, D. (1979b). An unusual free boundary problem from the theory of optimal harvesting. *Lectures Math. Life Sci.* **12**, 173–209.
Ludwig, D. (1980). Harvesting strategies for a randomly fluctuating population. *J. Cons. Int. Explor. Mer.* **39**, 168–174.
Ludwig, D., and Hilborn, R. (1983). Adaptive probing strategies for age structured fish stocks. *Canad. J. Fish. Aq. Sci.* **40**, 559–569.
Ludwig, D., and Varah, J. M. (1979). Optimal harvesting of a randomly fluctuating resource. II: Numerical methods and results. *SIAM J. Appl. Math.* **37**, 185–205.
Ludwig, D., and Walters, C. J. (1981). Measurement errors and uncertainty in parameter estimates for stock and recruitment. *Canad. J. Fish. Aq. Sci.* **38**, 711–720.
Ludwig, D., and Walters, C. J. (1982). Optimal harvesting with imprecise parameter estimates. *Ecol. Model.* **14**, 273–292.
Mangel, M. (1982a). Aggregation and fishery dynamics: multiple time scales, times to extinction, and random environments. *Ecol. Model.* **15**, 191–209.
Mangel, M. (1982b). Search effort and catch rates in fisheries. *European J. Oper. Res.* **11**, 361–366.
Mangel, M. (1984). "Search Theory." Springer-Verlag, Berlin and New York.
Mangel, M., and Beder, J. H. (1983a). Search and stock depletion theory and applications. *Canad. J. Fish. Aq. Sci.*, submitted.
Mangel, M., and Beder, J. H. (1983b). "Determining the Length of Fishing Seasons with Search Data." Technical report, Department of Mathematics, University of California, Davis, California.
Mangel, M., and Clark, C. W. (1983a). "Optimal Allocation of Searching Effort among Independently Fluctuating Fish Stocks." Technical report, Department of Mathematics, University of California, Davis, California.
Mangel, M., and Clark, C. W. (1983b). Uncertainty, search, and information in fisheries. *J. Int. Council Explor. Mer.* **41**, 93–103.
Mangel, M., and Plant, R. E. (1983). Multiseasonal management of an agricultural pest. I: Development of the theory. *Ecol. Model.* **20**, 1–19.

Martz, H. F., and Waller, R. A. (1982). "Bayesian Reliability Analysis." Wiley New York.

Matkowsky, B. J., Schuss, Z., Knessl, C., Tier, C., and Mangel, M. (1984). Asymptotic solution of the Kramers–Moyal equation and first passage times for Markov jump processes. *Phys. Rev. A* **29**, 3359–3369.

May, R. M. (1973). "Stability and Complexity in Model Ecosystems." Princeton Univ. Press, Princeton, New Jersey.

May, R. M., Beddington, J. R., Horwood, J. W., and Shepherd, J. G. (1978). Exploiting natural populations in an uncertain world. *Math. Biosci.* **42**, 219–252.

Maynard Smith, J. (1968). "Mathematical Ideas in Biology." Cambridge Univ. Press, London and New York.

Maynard Smith, J. (1974). "Models in Ecology." Cambridge Univ. Press, London and New York.

Menard, H. W., and Sharman, G. (1975). Scientific uses of random drilling models. *Science* **190**, 337–343.

Mendelssohn, R. (1978). The effects of grid size and approximation techniques on the solutions of Markov decision processes. Report 20H, Southwest Fisheries Center, National Marine Fisheries Service, Honolulu, Hawaii.

Mendelssohn, R. (1979). Determining the best trade-off between expected economic return and the risk of undesirable events when managing a randomly varying population. *J. Fish. Res. Board Canada* **36**, 939–947.

Mendelssohn, R. (1980). A systematic approach to determining mean–variance trade-offs when managing randomly varying populations. *Math. Biosci.* **50**, 75–84.

Merton, R. C. (1971). Optimal consumption and portfolio rules in a continuous time model. *J. Econom. Theory* **3**, 373–413.

Moffitt, L. J., and Farnsworth, R. L. (1981). Bioeconomic analysis of pesticide demand. *Agric. Econom. Res.* **33**, 12–18.

Mollison, D. (1977). Spatial contact models for ecological and epidemic spread. *J. Roy. Statist. Soc. Ser. B* **39**, 283–326.

Morse, P. M. (1977). "In at the Beginnings: A Physicist's Life." MIT Press, Cambridge, Massachusetts.

Moyal, J. E. (1949). Stochastic processes and statistical physics. *J. Roy. Statist. Soc. Ser. B* **11**, 150–210.

Nayfeh, A. H. (1973). "Pertubation Methods." Wiley, New York.

Newendorp, P. D. (1975). "Decision Analysis for Petroleum Exploration." Petroleum Publishing Company, Tulsa, Oklahoma.

Norman, J. M., and White, D. J. (1968). A method for approximate solutions to stochastic dynamic programming problems using expectations. *Oper. Res.* **16**, 296–306.

Ortega, J. M. (1972). "Numerical Analysis—A Second Course." Academic Press, New York.

Pella, J. J., and Tomlinson, P. K. (1969). A generalized stock production model. *Bull. Int. Amer. Trop. Tuna Comm.* **13**, 419–496.

Perkins, J. H. (1982). "Insects, Experts, and the Insecticide Crises." Plenum, New York.

Peterson, F. M., and Fisher, A. C. (1977). The exploitation of extractive resources: a survey. *Econom. J.* **87**, 681–721.

Pielou, E. C. (1977). "Mathematical Ecology." Wiley, New York.

Pindyck, R. S. (1978). The optimal exploration and production of nonrenewable resources. *J. Polit. Econom.* **86**, 841–861.

Pindyck, R. S. (1980). Uncertainty and exhaustible resource markets. *J. Polit. Econom.* **88**, 1203–1225.

Pindyck, R. S. (1981). The optimal production of an exhaustible resource when price is exogenous and stochastic. *Scand. J. Econom.* **83**, 279–288.

Pindyck, R. S. (1982). Adjustment costs, uncertainty, and the behavior of the firm. *Amer. Econom. Rev.* **72**, 415–427.
Pindyck, R. S. (1984). "Uncertainty in the Theory of Renewable Resource Markets." *Rev. Econ. Studies* **51**: 289–303.
Plant, R. E., Mangel, M. and Flynn, L. (1984). Multiseasonal management of an agricultural pest. II: Economic optimization. *J. Env. Econom. Man.*, in press.
Pratt, J. W. (1964). Risk aversion in the small and in the large. *Econometrica* **32**, 122–136.
Ras, N. J., Borwankar, J. D., and Rambrishna, D. (1974). Numerical solution of Ito integral equations. *SIAM J. Control* **12**, 124–139.
Reed, W. (1974). A stochastic model for the economic management of a renewable resource. *Math. Biosci.* **22**, 313–337.
Ricker, W. E. (1975). "Computation and Interpretation of Biological Statistics of Fish Populations." Bulletin 191. Fisheries Research Board of Canada, Ottawa, Ontario.
Riordan, J. (1979). "Combinatorial Identities." R. E. Krieger, Huntington, New York.
Ripley, B. D. (1977). Modelling spatial patterns. *Proc. Roy. Statist. Soc. Ser. B* **39**, 172–212.
Ripley, B. D. (1981). "Spatial Statistics." Wiley, New York.
Ross, S. M. (1970). "Applied Probability Models with Optimization Applications." Holden-Day, San Francisco, California.
Ross, S. M. (1980). "Introduction to Probability Models." Academic Press, New York.
Ross, S. M. (1983). "Introduction to Stochastic Dynamic Programming." Academic Press, New York.
Rothschild, B. J. (1977) Fishing effort. *In* "Fish Population Dynamics" (J. A. Gulland, ed.), pp. 96–115. Wiley, New York.
Rothschild, B. J., and Suda, A. (1977). Population dynamics of tuna. *In* "Fish Population Dynamics" (J. A. Gulland, ed.), pp. 309–334. Wiley, New York.
Roughgarden, J. (1979). "Theory of Population Genetics and Evolutionary Ecology: An Introduction." Macmillan, New York.
Rubinstein, L. I. (1971). "The Stefan Problem." American Mathematical Society, Providence, Rhode Island.
Sage, A. P. (1977). "Systems Engineering: Methodology and Applications." IEEE Press, New York.
Sage, A. P., and Melsa, J. (1980). "Estimation Theory with Applications to Communications and Control." R. E. Krieger, Huntington, New York.
Samarskii, A. A. (1963). On an economical method for the solution of a multidimensional parabolic equation in an arbitrary region. *U.S.S.R. Comput. Meth. Math. Phys.* **2**, 896–926.
Schaefer, M. B. (1957). Some considerations of population dynamics and economics in relations to the management of marine fisheries. *J. Fish. Res. Board Canada* **14**, 669–681.
Schoemaker, P. J. H. (1982). The expected utility model: its variants, purposes, evidence and limitations. *J. Econom. Lit.* **20**, 529–563.
Schuss, Z. (1980). "Theory and Application of Stochastic Differential Equations." Wiley, New York.
Seber, G. A. F. (1982). "The Estimation of Animal Abundance." Macmillan, New York.
Sengupta, J. K. (1982). "Decision Models in Stochastic Programming." North-Holland Publ., Amsterdam.
Silvert, W. (1977). The price of knowledge: fisheries management as a reasearch tool. *J. Fish. Res. Bd. Canada* **35**, 208–212.
Smiley, A. K. (1979). "Competitive Bidding Under Uncertainty." Ballinger, New York.
Smith, A. D. M., and Walters, C. J. (1981). Adaptive management of stock recruitment systems. *Can. J. Fish. Aq. Sci.* **38**, 690–703.

Smith, C. W. (1976). Option pricing. *J. Finan. Econom.* **3**, 3–51.
Smith, V. K., and Krutilla, J. V., eds. (1982). "Explorations in Natural Resource Economics." Johns Hopkins Press, Baltimore, Maryland.
Stoer, J., and Bulirsch, R. (1980). "Introduction to Numerical Analysis." Springer-Verlag, Berlin and New York.
Stefanou, S. (1981). "The Competitive Industry Extraction Profile When the Tax Policy Is Uncertain: A Model for Exhaustible Resources." Unpublished report, Department of Agricultural Economics, University of California, Davis, California.
Stone, L. D. (1975). "Theory of Optimal Search." Academic Press, New York.
Sukhatme, P. V., and Sukhatme, B. V. (1970). "Sampling Theory of Surveys With Applications." Iowa State University Press, Ames, Iowa.
Swierzbinski, J. (1984). Statistical methods applicable to selected problems in fisheries biology and economics. *Mar. Res. Econom.* **1**, forthcoming.
Swierzbinski, J., and Cain, K. C. (1981). "The Choice of Sampling Scheme for Population Counts in a Patchy Environment and Its Consequences for the Cost and Precision of Estimates." Unpublished report, Department of Applied Mathematics, Harvard University, Cambridge, Massachusetts.
Szidarovsky, F., and Yakowitz, S. (1978). "Principles and Procedures of Numerical Analysis." Plenum, New York.
Taylor, C. E., and Georghiou, G. P. (1979). Suppression of insecticide resistance by alteration of gene dominance and migration. *J. Econom. Entomol.* **72**, 105–109.
Taylor, C. E., and Georghiou, G. P. (1982). Influence of pesticide persistence in evolution of resistance. *Env. Entomol.* **11**, 746–750.
Thom. R. (1975). "Structural Stability and Morphogenesis." Benjamin, New York.
Turelli, M. (1977). Random environments and stochastic calculus. *Theoret. Population Biol.* **12**, 140–178.
Uhlenbeck, G. E., and Ornstein, L. S. (1930). On the theory of Brownian motion. *Phys. Rev.* **36**, 823–841.
Uhler, R. S., and Bradley, P. G. (1970). A stochastic model for determining the economic prospects of petroleum exploration over large regions. *J. Amer. Statist. Assoc.* **65**, 623–630.
Vadja, S. (1970). Stochastic programming. *In* "Integer and Nonlinear Programming" (J. Abadie, ed.). North-Holland Publ., Amsterdam.
Van Kampen, N. G. (1981). "Stochastic Processes in Physics and Chemistry." North-Holland Publ., Amsterdam.
Van Mellaert, L., and Dorato, P. (1972). Numerical solution of an optimal control problem with a probability criteron. *IEEE Trans. Automat. Control* **AC17**, 543–546.
Varley, G. C., Gradwell, G. R., and Hassell, M. P. (1973). "Insect Population Ecology." University of California Press, Berkeley, California.
Wald, A. (1973). "Sequential Analysis." Dover, New York.
Walters, C. J. (1981). Optimum escapments in the face of alternative recuitment hypotheses. *Canad. J. Fish. Aq. Sci.* **38**, 678–689.
Walters, C. J., and Hilborn, R. (1976) Adaptive control of fishing systems. *J. Fish. Res. Board Canada* **33**, 145–159.
Walters, C. J., and Ludwig, D. (1980). Effects of measurement errors on the assessment of stock recruitment relationships. *Canad. J. Fish. Aq. Sci.* **38**, 704–710.
Waltman, P. (1983). "Completion Models in Population Biology." SIAM, Philadelphia, Pennsylvania.
Washburn, A. R. (1981). "Search and Detection." Millitary Applications Section, Operations Research Society of America, c/o Ketron, Inc., Arlington, Virginia.

REFERENCES

Watson, T. F., Moore, L., and Ware, G. W. (1975). "Practical Insect Pest Management." Freeman, San Francisco, California.
Wax, N. (1954). "Selected Papers on Noise and Stochastic Processes." Dover, New York.
Weitzman, M. L. (1976). The optimal development of resource pools. *J. Econom. Theory* **12**, 351–364.
Wenk, C. J., and Bar-Shalom, Y. (1980). A multiple model adaptive dual control algorithm for stochastic systems with unknown parameters. *IEEE Trans Automat. Control* **AC-25**, 703–710.
White C., and Kim, K. W. (1980). Two solution procedures for solving vector criterion Markov decision processes. *Large Scale Syst.* **1**, 129–140.
Wilson, L. T., et al. (1983). Within plant distribution of spider mites (*Acari Tetranychidae*) on cotton: A developing implementable monitoring program. *Env. Entomol.* **12**, 128–134.
Wong, E. (1971). "Stochastic Processes in Information and Dynamical Systems." McGraw-Hill, New York.
Wonham, W. M. (1969). Random differential equations in control theory. *In* "Probabilistic Methods in Applied Mathematics." vol II. (A. T. Bharucha-Reid, ed.), pp. 131–212. Academic Press, New York.
Yakowitz, S. (1969). "Mathematics of Adaptive Control Processes." Elsevier, New York.
Young, P. (1978). General theory of modeling for badly defined systems. *In* "Modeling, Identification, and Control in Environmental Systems." (L. Vansteenkiste, ed.), pp. 103–135. North-Holland Publ., Amsterdam.
Zacks, S. (1971). "The Theory of Statistical Inference." Wiley, New York.

Index

A

Adaptive management, 155, 170, 188
Adjoint variable, 82, 109, 215
Age-structured models
  fisheries, 188
  pest control, 206
Aggregating fisheries, 148
Aitken's method, 226
Algorithm
  Charnes–Copper for search, 100
  filtering, 36
  first-order, mean and variance propagation, 31
  second-order, mean and variance propagation, 31
Asymptotic expansions, 160, 179

B

Backward equation, 20
Backward induction, 58
Bang-bang control, 82, 179
Bayesian analysis and approaches, 54, 116, 165
Bayes theorem, 11, 33, 117
Bernoulli process, 16
Beta density, 57
Bifurcation, 153
Binomial distribution, 16, 57, 98, 166
Boundary conditions, 110, 207, 234

Brownian motion, 17, 28, 44, 128
  numerical computations involving, 229

C

Calculus of variations, 80
Capture of organisms in sampling, 139
Carrying capacity, 149, 185
Catastrophe, 154
Catch per unit effort (CPUE), 152
Central limit theorem, 162
Chapman–Kolmogorov relations, 20
Characteristic function, 15
Characteristics, 25
Charnes–Cooper algorithm, 100
Coefficient of variation, 7, 146
  minimizing, as an optimization criterion, 141
Combinatorial identities, 168
Common property resource, 1
Concavity, 109
Conditional density, 10, 71
  for detection in transect theory, 134
Conditional probability, 10
  of detection in search, 97
Conjugate prior, 54
Constraints
  optimal control problems, 89
  resource extraction, 143
Consumption rate, path or profile, 81, 118
Contagious distributions
  negative binomial as a model for, 96

Control
  bang-bang, 82
  deterministic, 33, 116, 118
  feedback, 58
  with probability criteria, 61
  stochastic, 33, 39
Correlation function, 18
Costs
  in control functional, 39
  extracting resources, 81
  population enhancement, 63
  production of resources, 196
  search, 148
  transects, 137
Cotton, 204
Crop–pest interactions, 214

D

Decision rules, 69
Delta function, 7, 18
Density estimation
  in transect theory, 133
Density function
  beta density, 57
  discrete, as a model for resource deposits, 117
  exponential, 33, 160
  gamma density, 8, 10
  log-normal, 93
  normal density, 7, 17
  posterior density, 54, 105
  Rayleigh density, 66
  steady-state, 38, 196
Detection, *see also* search
  capability, 158
  probability, 97, 158
  width (sweep width), 133
Deterministic control, 82, 105
  as a limit of stochastic control problem, 52, 127
Deterministic dynamic programming, 42, 219
Deterministic limit, 52
Difference equation, 23, 46, 219
Differential-difference equation, 122
Differential equation, 49, 60, 97, 177, 230
  solutions obtained by guessing the appropriate form, 46
Diffusion equation, 19, 197
Diffusion limit, 52, 127
Diffusion process, 19

Discount rate, 48, 218
Distribution function, 6
  binomial distribution, 16
  exponential distribution, 33
  negative binomial distribution, 12
  normal distribution, 6
  Poisson distribution, 12
Dynamic programming
  deterministic, 42, 110
  stochastic, 44, 50, 53, 110, 122, 128, 186, 194, 226
Dynamic programming equations
  Brownian motion, 44, 128, 176, 186, 194, 234
  jump processes, 50, 122
  Myopic Bayes approach, 65
  parameter uncertainty, 53, 110
  solutions obtained by guessing the appropriate form, 46

E

Effort allocation
  resource exploitation, 100, 174
  search, 101, 141
End conditions, 177
  for adjoint variable in optimal control, 90, 216
  in objective functionals, 33
End times
  unknown in optimal control, 108
Equilibrium
  population level, 86
  probability density, *see* Steady-state probability density
  stock level, 86
  price–stock equilibria, 87, 196
Escapement, 157, 170
  as an optimization parameter, 173
Estimates
  of density in transect theory, 134, 145
  in Kalman filtering, 33
  maximum likelihood, 97
Euler equations, 81, 88
Exhaustive search, 98
Exit times, 21, 62
  as control objectives, 62, 200
  defined, 21
Expectation, 7, 26, 27, 57
  conditional, 11, 71
  involving nonlinear functions, 39, 113

INDEX

Exploration, 94, 122, 130
Exponential density, 33, 160
Extraction rate, path or profile, 81, 118
Extremizing path, 88

F

False detections, 158
Feedback controls, 58
First-order condition, 109
Fisher information, 136, 165
Fishing effort, 151
Fishing rights, 170
Fixed points
   in Newton's method, 222
Fokker–Planck equation, 21, 196
Forest growth and management, 40
Forward equation, 21
Free boundary problems, 125

G

Gamma distribution, 8, 68
Gamma function, 8, 142
Gaussian white noise, 18
Generating function, 15, 23
Genetic models
   in pesticide resistance management, 211
Growth rate, 75
   intrinsic, 151, 207
   with pesticide applied, 207

H

Hamiltonian, 82
Hardy–Weinberg formula, 212
Harvest, 161
   expected, with parameter uncertainty, 26, 166, 193
   quotas, 171
   strategies, 186
Helium controversy, 94, 203
Heterozygous, 211
Homozygous, 211

I

Independent increments, 17
Information, 54
Initial distribution, 33
Integrability condition, 180

Integrodifferential equations, 110
Ito calculus, 17, 45, 129

J

Jacobian, 225
Joint density, 8
   marginal density defined from, 8
Jump process, 22, 28, 50

K

Kalman filtering, 33
   applied to forest management, 132
   linear, 36
   nonlinear, 38
Kuhn–Tucker theorem, 143

L

Lagrange multiplier, 80, 104
Lateral range curve, 133
Law of total probabilities, 11
Learning, 106
   effects on extraction rates, 108, 113
   by updating parameters, 53
Lethal dose, 205
Likelihood function, 132, 142, 162, 167, 190
Limits
   negative binomial and Poisson, 54
   Poisson of the binomial, 98
Line transect theory, 132
Linear regulator, 46
Local proportional risk aversion, 78
Logistic equation, 85
Log-normal distribution, 93

M

Marginal distribution, 8
Marginal economic return, 26
Maximum likelihood estimates, 97, 135, 165
Maximum sustainable yield (MSY), 186
Mean, *see also* Expectation
   propagation in stochastic differential equations, 27
Mean–variance models, 33, 198
Measurement, 34
   errors, 138, 189
   noisy, 35
Method of characteristics, 25, 208

Method of lines, 235
Mineral resources, 93
Model selection
  and adaptive management, 156, 188
  for aggregating fisheries, 155
Monopolist, 130
Myopic Bayes approach, 65

N

Negative binomial distribution, 12, 54, 94, 141
Newton's method, 191, 221
Nonlinear equations, 114, 202, 217
  algebraic, solved by Newton's method, 221ff
  differential, 30, 49
Nonlinear least squares, 155
Nonlinear measurements, 39
Nonlinear programming, 65, 100, 141
Normal distribution, 7, 9, 17, 38, 120

O

Objective functionals, 45, 76
Optimal controls
  deterministic, 33, 116, 118, 215
  feedback, 58
  stochastic, 41
  stationary, 72
Optimal effort levels, 141
Ornstein–Uhlenbeck process, 38, 63
Orthogonal polynomials, 134

P

Parameter estimation, 155
  accuracy of, 193
Parameter optimization problems, 215
Parameter uncertainty, 188, 192
Pareto optimal control, 199
Partial differential equations, 19, 23
  for age structured populations, 206
Patchiness, 95
  effects on survey design, 145
Patchy encounters, 141
Pattern recognition, 121
Performance index, 46
Pest control, 203
Poisson distribution or process, 12, 22, 94, 122, 141
  extended to include depletion, 164
Policy iteration, 72
Pontryagin maximum principle, 82, 105
Population enhancement, 63
Porpoise, 148
Portfolio selection, 48
Posterior density, 54, 106
Posterior distribution, 54, 167
Precision, 7, 146
Preposterior expectation, 54
Present value, 128
Price, 81
  fluctuating, 127
  in objective functionals, 127, 194
Price dynamics, 81, 127, 129, 195
Principle of optimality, 43, 118
  in stochastic control, 43
Prior density, 105
  improper, 167
  informative, 54
  noninformative, 54
Probabilistic programming, 61, 198
Probability criterion, 63
Profit, 128, 171
  in objective functionals, 125
Profits tax, 83
Pure birth process, 23

R

Random encounters, 97, 141
Random search formula, 98
Random variable, 6
  quotients of, 10
  sums of, 8
Rayleigh density, 66
Recruitment, 171
Regression methods, 190
Remaining reserves, 117
  in resource exploitation, 112
Renewal processes, 159
  applied to fisheries, 158ff
  cental limit theorem, 162
  defined, 158
  main renewal theorem, 161
Reserves, 128
Resistance to pesticides, 205
Resources
  classified, 74
  exhaustible, 79, 93ff
  layered, exploitation of, 112

INDEX

mixed, 203
renewable, 85, 132ff
Ricatti equation, 125
Ricker curve, 157, 189
  stochastic version of, 178
Risk aversion, 77
Risk neutral, 77
Risk premium, 77
Runge–Kutta method, 231

S

Sample path, 17
Schooling rate, 151
Search
  exhaustive search, 98, 133
  random search, 24
  for resource deposits, 98
  for schools of fish, 4, 24, 53, 158
  with depletion, as an extension of the Poisson process, 24, 164
Semi-Markov process, 22
Sequence and series acceleration methods, 226
Sequential analysis, 121
Shadow price, 82
Sighting surveys, 34
Singular path, 83
Singular perturbation methods and problems, 179
Splines, 236
Spraying of pesticides, 208
Stationary controls, 72
Steady-state probability density, 38, 196
Steady states of differential equations, 38
Stochastic control, 44ff, 58
Stochastic differential equation, 28, 44, 61, 176, 194, 208
  mean and variance propagation, 27
  numerical solutions, 228
Stochastic dynamic programming, 44, 50, 53, 110, 122, 128, 179, 186, 194, 226
Stock abundance, 158
Stock dynamics
  Beverton–Holt model, 185
  logistic model, 86, 150, 185
  Pella–Tomlinson model, 185
  Ricker model, 157, 171
Stock–recruitment curves, 171
Stock surveys, 138
Stopping times, 102

Strategies, 68, 186
Susceptibility to pesticides, 203
  as an exhaustible resource, 206
Sweep width, 133
Switching points of effort, 180, 187
  and singular path, 180
Systems engineering, 4

T

Taxes, 83
Taylor expansions, 6, 17, 39, 41, 45, 98
  involving Brownian motion, *see* Ito calculus
Terminal values, 177
Towed-net samplers, 138
Transition density, 19
Transversality condition, 108
Two-armed bandit problem, 56
Two-point boundary value problem, 90

U

Unbiased estimate, 136, 146
Update measurement, 33
Updating, 54, 116
Utility function, 76
  linear, 127
  logarithmic, 109, 115
  power, 76, 109, 115
  quadratic, 123
  and risk behavior, 77
  vector-valued, 78
Utility of consumption, 112

V

Value function, 68
Variance, 7, 17, 27, 40, 57, 116, 178
  in transect estimates, 137
Variation, 7, 146

W

Waiting time, 22
Weiner filtering, 33

Y

Yield–effort curves, 150